Avidly Reads SCREEN TIME

Avidly Reads

General Editor: Sarah Mesle

Founding Editors: Sarah Blackwood and Sarah Mesle

The Avidly Reads series presents brief books about how culture makes us feel. We invite readers and writers to indulge feelings—and to tell their stories—in the idiom that distinguishes the best conversations about culture.

Avidly Reads Theory
Jordan Alexander Stein

Avidly Reads Board Games
Eric Thurm

Avidly Reads Making Out
Kathryn Bond Stockton

Avidly Reads Passages
Michelle D. Commander

Avidly Reads Guilty Pleasures
Arielle Zibrak

Avidly Reads Opera
Alison Kinney

Avidly Reads Poetry
Jacquelyn Ardam

Avidly Reads Screen Time
Phillip Maciak

Avidly Reads

Screen Time

PHILLIP MACIAK

NEW YORK UNIVERSITY PRESS *New York*

NEW YORK UNIVERSITY PRESS
New York
www.nyupress.org

Library of Congress Cataloging-in-Publication Data
Names: Maciak, Phillip, author.
Title: Avidly reads screen time / Phillip Maciak.
Description: New York : New York University Press, [2023] | Series: Avidly reads | Includes bibliographical references and index.
Identifiers: LCCN 2022049832 | ISBN 9781479820542 (hardback) | ISBN 9781479820573 (paperback) | ISBN 9781479820580 (ebook) | ISBN 9781479820597 (ebook other)
Subjects: LCSH: Television viewers. | Television—Social aspects. | Information technology—Social aspects. | Digital media—Social aspects.
Classification: LCC HE8700.65 .M33 2023 | DDC 302.23/45—dc23/eng/20221021
LC record available at https://lccn.loc.gov/2022049832

New York University Press books are printed on acid-free paper, and their binding materials are chosen for strength and durability. We strive to use environmentally responsible suppliers and materials to the greatest extent possible in publishing our books.

Manufactured in the United States of America

10 9 8 7 6 5 4 3 2 1

Also available as an ebook

To Maeve, Phoebe, and Mel,

my real life

Contents

Figure 1.1. *Mad Men*, "Waterloo," 2014

1

CREATURES MADE OF SCREENS

We are the TV. They're looking at us. It's season seven of *Mad Men,* and the Moon landing is being broadcast live on television. They're captivated, they're in awe. They're warmed by each other's presence. And they're illuminated by the screen in front of them.

For a brief moment, the camera lets us be the objects of their gaze. We see what it's like to be *watched* like a screen, like Neil Armstrong, like a commercial for hamburgers or floor wax. And we look back at them in a way that Neil Armstrong can't, but maybe some other screens can: we see Peggy Olson posed upright in a baby blue dress, Don Draper lurched forward in rumpled shirt and tie. They, as viewers, become *our* object. Attentive, but vulnerable. Looking out, looking in.

Michel Foucault, a contemporary of Don Draper and Peggy Olson, wrote once about a similar figure in a similar image, this one by Velázquez: "He is staring at a point to which, even though it is invisible, we, the spectators, can easily assign an object, since it is we, ourselves, who are that point: our bodies, our faces, our eyes." We are the TV. They're looking at us. We, ourselves, are that point.

This image has always stuck with me, a viewer and a writer for whom these people on this TV show were uncommonly important, and someone whose life is shaped by screens of all kinds. This frozen moment of two spectators whose spectatorship, and its relationship to our own, isn't some pause in the real story. It's *the* story.

The screen watching us watching. All the time. This is the story of our lives, as workers, as consumers, as friends, as lovers, as parents, partners, children. Screens mediating whatever it is we do, back and forth, in and out. Maybe the image tells a story about communion and community and innovation across time and space. Our far-flung relatives can be close to us despite their physical distance, information that would otherwise have been remote is accessible within seconds, kids can go to school without exposing themselves to life-threatening illness, new voices can express themselves through expanded, if not fully democratized, online platforms. Or maybe the story reveals surveillance and exploitation and vulnerability. Virtual closeness erodes the importance of real community, the slippery ease of the internet glosses over its massive carbon footprint, online spaces of connection are also spaces of violent radicalization, bullying, and hate, everybody is watching you.

That image's looping self-awareness disguises the fact of its incredible straightforwardness. The hope and paranoia indexed by that screenshot of Peggy and

Don aren't a televisual fantasy—those feelings are ordinary, everyday. It's euphoric and it's depressing, it's both our favorite show and an existential nightmare of complicity and exposure. This is what it's like. This is how it feels. *This*—the charged, sometimes confused, space between these people, their hot screen, and us—is what's happening. It's happening now.

What's happening is *screen time*.

This is a book about how screen time—the time you spend looking at your screen or the time the people you love spend looking at theirs—makes us all feel. For a lot of people, that's not a very complicated topic. Screen time makes us feel bad. For as long as that phrase has been a part of our lexicon, it's been an avatar for guilt, for shame, for regret. It's bad for our children, it's bad for their eyeballs, it's bad for America. And not only that, but there's way too much of it. Nobody feels proud of their screen time. Nobody wants to increase their screen time. Screen time makes us feel bad. By design.

Which makes it interesting that, since 2018, Screen Time has also been an app. Starting with iOS 12 in 2018, Apple made this feel-bad function a part of its actual screen design. The Screen Time app, then, is also a screen watching us watching, just as much as the image of Peggy and Don and the Moon landing. Is there something this app can teach us about a way that, alongside the badness, screen time can make us feel . . . good? Or, maybe not good, but something at least a little more complicated than "bad."

The app, which comes built in to all Apple products now, can do two things. The first thing it can do is tell you about yourself. It tracks, with your permission, everything you look at on your iPhone and for how long. It clumps these things into three categories—Social, Entertainment, and Productivity & Finance—and tells you how much you dip into these categories every day and every week. It comes up with a personal average to weigh yourself against. Perhaps most importantly, it alerts you to deviations from the norm, whether you exceed your personal average screen time or dip below it. It doesn't, mind you, say in any proscriptive way what a healthy amount of screen time might be—*this much* screen time is *too much* screen time. It simply quantifies your own relationship to the screen, in a weirdly moral way. You could spend a paltry thirty minutes a day looking at your phone, and Screen Time would still be able to give you a sting of guilt if you hit forty-five or a charge of victory if you make it to fifteen. But it also doesn't really want you to stop. You'll still get a pat on the back if you bring your daily usage down from sixteen to fifteen hours.

The other thing Screen Time can do is suggest that controlling your screen time is something you could and should do. It tells you whether or not you are in control, and it gives you tools to regulate your own screen time, so as to become more in control. You can set limits for particular apps or particular categories—only an hour of Twitter per day, no email after hours. You can restrict content for yourself or for your children. You can be the screen watching others watch. Every look is doing something, asserting something, and you can be in control of that action. (Foucault had something to say about this, too.)

What Screen Time tells you, implicitly, is that all of this looking that this device enables is bad for you in excess. But you didn't really need an app to know that, did you? If you've grown up in the US in the past thirty years, you know, very instinctively if also confusedly, that screens are bad for you. You know that you need to control the time you devote to them. Even when they bring you sustained and meaningful pleasure—watching a Super Bowl or a big finale or scrolling family pictures on Instagram or texting your friends—you know, at some level, that there could be too much of this pleasure, and that that would lead to some inchoate ill. You have been told this your whole life. You know, further, that the way to interact with screens is through measurement, through surveillance, maybe even through abstinence. It's the way you encounter all screens: you know to be worried about your relationship to them. You know—even when you question that knowledge, even when you enjoy that knowledge—that they are a problem.

And that's because you yourself are integrated with that app in ways Apple was canny enough to realize. What you see on your phone is just a visualization, a cute externalization, of a process that's likely already going on and on inside your own mind, whether you buy the moral framing or not. It's a screen that gives you access to millions and millions of images and words and even people, and it pays attention to what you're doing when you hold it in your hand, and it judges you the way you might judge yourself. It's an autofill for an anxiety that was born with the television in the fifties and caught fire at the end of the twentieth century, creating a totalizing atmosphere for our lives. The screen sends you a notification. It asks you to look at it. And it makes you feel bad, or bad *and something more*, when you do. This is what it's like. This confusion is how it feels. This is screen time.

So this is a book about that feeling, where it comes from, what we can do with it, and what it does to us. The screen time era spreads from the early nineties to the present, and it's a time period defined by big changes in screen technology and in the way people talk about screen technology. These were popular national conversations, happening everywhere from local papers to TV talk shows, but they did not impact everybody equally. Screen time is collective, and it is also hyper-individualized. So this book about the culture of screen time is also, by necessity, about me. I grew up in the screen time moment, a white, cisgender, heterosexual boy in a suburban household in Western Pennsylvania with access to screens of all kinds. We had TVs, we had premium cable, we had home computers as early as the Apple II. I had a Nintendo, and later a Game Boy. My first iPhone was an iPhone 5. My first iPad was the first iPad. I watch TV now as a regular person, as a parent of two small girls, as a TV critic, as a media scholar, and as a college professor. This was my coming of age in screen time. No matter how similar it may seem, it was not yours.

And that's true for structural reasons as well as personal ones. A lot of the discussion in this book about screen time presupposes a relatively high level of availability. Access to technology, access to the internet, access to screens for virtual school or virtual work, access to the safety net of reliable Wi-Fi, affordable daycare, affordable elder care, or even the physical space required for those screens to matter—these are considerably bigger social problems than, say, my personal ambivalent feelings about binge-watching. But screen time is personal, and so this book—even as it's built around some claims about the way that the popular discourse of screen time has impacted all of us culturally since the nineties—finds examples in my own screen time. Even as I'll sometimes speak about how screens address us universally, I do so with the knowledge that my own experience of those screens is not universal. Some of this book you'll recognize; other parts you won't.

This is my screen time, and it starts now.

* * *

When I was in elementary school in the early 1990s, before it was an app, the phrase "screen time" emerged as a meme to scare parents about the dangers of Too Much TV for little kids. The term, in its current form, originates in a 1991 *Mother Jones* article by the opinion columnist Tom Engelhardt. Previously, "screen time" had referred to how much time an actor appeared onscreen in TV and movies. But Engelhardt, in "The Primal Screen," reversed the term's meaning. Screen time wasn't a measure of what happened *on* the screen; it was a metric evaluating *us*.

And as such, it was sociological, rather than aesthetic—something to be fretted over, policed, and blamed, a low-grade moral panic. Engelhardt's essay made an argument about how kids should look at screens, what screens meant in their lives, and why we—as parents, or just as responsible adults—should regulate their exposure to them. But it was also an argument to feel a certain way about screens, to feel bad, to feel skeptical, to feel animosity or antagonism or even fear toward them. It was an act of cultural definition.

In the intervening decades, that definition has become definitive. For parents, guesstimating and regulating kids' screen time is now a huge part of the job. Whether taking a hardline or agnostic position, it's become a central facet of modern childrearing, a choice like deciding whether to raise kids religious or when to allow them to get their ears pierced. How much is too much? What are they watching when I'm not paying attention? What might they see? Who might see them? We worry about what our kids watch; we worry about what might be in our screens watching them.

As the screens Engelhardt worried about have only become more intimate, more integrated, and more seemingly invasive, the panic about them has remained at a low boil. This is why screen time feels so bad. No longer a discrete concern about kids and TV, it hangs over everything, impacting the way people watch and the way people make things for other people to watch. Engelhardt's regulatory impulse and the bad vibes that have followed it around have shaped everything for the past thirty years. Here in the early parts of the twenty-first century, screen time is a metric for everybody.

But what are we measuring? If even the screen in my pocket knows about screen time, what does it know? Over the course of a day, one single screen—let's say the smudgy screen of that iPhone in my pocket—plays host to dozens of little rituals and compulsions. It might be the interface I use to watch an episode of TV, watch a viral YouTube video, attend a work meeting, make a bank deposit, write a few sentences for a creative project, write an email, take a picture of my daughter and post it on Instagram, "like" peoples' pictures of their children on Instagram, "like" somebody's post about getting sick on Instagram, "like" a picture comparing a screengrab of a basketball game to a work of modern art, tell my mother I love her, read a news story about the pandemic, watch an unedited video of a natural disaster, spend hours (split up into thirty-second intervals) scrolling Twitter, read a review of a new restaurant, look at pictures of food from that restaurant on the chef's Instagram account, text my partner Mel about whether she's in the mood for the cuisine that restaurant serves, order takeout from that restaurant, find precise directions to the takeout restaurant, watch a live update of regional weather radar, change the temperature in my house, laugh until I cry at a tweet about a TV show, use an app to identify an insect, use an app to identify a piece of space junk slowly, but perceptibly floating over the place where I live.

This isn't meant to be an arch litany of consumer culture's many poisonous tendrils. It's just a fairly rigorous audit of what I did—as of this writing—last Friday. It's easy to moralize, to read this list and see merely another zombie staring at a screen all day.

It's also easy to imagine that all of these activities are separable. Social life, work life, family life, the life of a consumer or a spectator or a stargazer. But that separation is a convenient fantasy. It's imaginary. All these things are happening at once in the same place in the same screen time.

And all of those things on that screen—and on all the other screens—meet up in my head. They bump into each other, share space, and eventually begin to reflect and reenvision each other. The same way that television is inseparable from the people we watch it with and the places where we watch it, the individual screens themselves all become part of the same pulsing archive. What would happen if we took that jumble seriously? What might we learn about the kinds of choices all of us make when presented with that dynamic, algorithmic, aesthetic temporality?

Tom Engelhardt, in 1991, would have been terrified about all of this. (He might still be now!) But fascinatingly, the statistics Engelhardt cites about average daily screen time for children are roughly similar to such statistics thirty years later, or at least not so dramatically different as you might assume. What has substantially risen over that time isn't the amount of time spent around screens, it's the number of screens around which we—all—spend our time.

So it's important to remember that the child about whom Engelhardt was concerned was not one juggling personal screen devices like an overworked executive with no time for love at the beginning of a romantic comedy, but rather a child-shaped lump, sitting, inert, two feet away from a gigantic, throbbing television set. The screen that worried Engelhardt—and the one that shaped our sense of what we, too, should be worrying about—isn't the screen that most of us spend our time with now. This is key for the feelings that surround screen time: the particular material, creative, and commercial structures of television gave form to an anxiety that attaches, in all sorts of strange ways, to a range of other devices.

For purposes of family time, social life, and even interior design, the television was an organizing feature of the home. And it's that *organization* that worries Engelhardt most. Engelhardt keeps repeating this word: screen time is "an organizing principle for children's experience"; it's a feature of the "corporate organization of childhood." This was not a new worry; an influential 1961 study called *Television in the Lives of Our Children*, which was mostly sanguine about the effects of TV, singled out the *organization* of time as television's biggest threat to little ones in much the same way Engelhardt does. And this corrupting organization of what would presumably have been innocently *dis*organized time otherwise, this locking of kids' looks on the many screens of the Clinton era home, is most beneficial to the corporations who are staring back.

"Children's out-of-sight time was now being reorganized as screen time," Engelhardt writes, "and opened up to the gaze of adults; and the adults doing the gazing had something in mind other than instructing or parenting. For those linked screens had pulled children into a different universe, with a new kind of time and new kinds of desires." Engelhardt's *Mother Jones* article rested on a light, left, anti-consumerist critique that seems almost tangential to contemporary arguments about screen time. Indeed, of Engelhardt's many salient points about consumer surveillance and invasion, what we seem to have kept are the familiar elements of moral panic. In other words, screen time, in Engelhardt's rhetorical formulation here, seems unmistakably like a sex thing.

These business interests putting *Scooby-Doo* on your TV are not just financial concerns or corporate entities for Engelhardt; they are "adults" gazing at children. McDonald's isn't just trying to sell your child a Happy Meal. Ronald McDonald himself—that creep—is looking, *gazing,* at your pajama-clad child as they sit crisscross applesauce in front of the surveillance device you, the "out-of-sight" parent, brought into the living room. And Ronald and Scoob and all the other "adults doing the gazing" have "something in mind other than instructing or parenting"—fill in the first perverted gerund that pops into your head! And, more than that, more than just being gazed upon by multinational businesses, these children are alone with these personified corporations, pulled "into a different universe" the way they might be pulled into a van by a stranger. And what happens to these children? Gazed at, pulled, and who knows what else by Fortune 500 companies? Well, they are given "a new kind of time and new kinds of desires." Screen time is a time animated by the *new desires* of children. Ruh roh.

Another way of saying this is that the anxiety about "screen time" is about kids growing up. Or, rather, it's about kids growing up the wrong way—too fast, too slow, too independent, too unsupervised, too supervised by the wrong kind of supervision. Engelhardt even comes right out and says it, in terms somehow both vague and poignant. After listing off the scandalous numbers of households in possession of camcorders and cable boxes, Engelhardt writes, "Growing up is the process by which we protect ourselves against the situation such numbers describe." The mournfulness of this sentence hits me. What situation, exactly, do these numbers describe? It seems naïve to imagine that adults of Engelhardt's generation grew *out* of their imbrication in consumer culture rather than more and more beholden to it. Indeed, growing up, for Engelhardt, seems to entail less an escape from those tendrils than an increasing awareness—or lack of awareness—of where they grab us most powerfully. "Maturity," he writes, "means the ability to discern." And that ability to discern is eroded by what he perceives to be the increasing integration of storytelling and advertising, art and commercials.

Somewhat strangely, it's the era of Don and Peggy to which Engelhardt looks back most fondly, a mythical time when ads were ads, not content. It's not the media on their own terms that cause the greatest fear for Engelhardt, but their relationship to time, the vanishing distinction between the time kids are watching their favorite shows and the time they are watching advertisements. (This is somewhat of a misleading narrative, though, as television—especially children's television—has always been inextricable from advertising.) He ends the article praising *Mister Rogers' Neighborhood,* not just for its lack of commercial breaks, but for its slowness and its adherence to the "real-time boundaries" of the world Rogers and the viewers at home share. Rogers refuses to sync his show to the time signature of advertisement, to reshape time according to the new desires the adults in the TV wish to produce.

It's hard to deny this sanctification of the Neighborhood of Make-Believe. Rogers was a genius of television as a medium as well as a "genius of empathy," as Ashley C. Ford calls him. His understanding of the address television allowed, its capacity for quiet, its status as an art form *in time* are perhaps so extraordinary for how infrequently they've been replicated. Truly nobody makes television like Mister Rogers did; he provided a formal model for what to do with all that intimacy TV produces that precious few producers have ever chosen to follow. For Engelhardt, Rogers provided a reminder, taught an awareness, of the world outside the screens. His exceptionality is a bad portent.

The opposite of that is *Pee-wee's Playhouse*. For Engelhardt, Pee-wee, by way of contrast, was a creature *made* of screens. "Put a human being in such a screen world, he seems to say, and I will be the result," writes Engelhardt near the end of the article. "The speedier the screens, the more inert, ultimately, the viewers." We, then, are Engelhardt's creatures made of screens. We are the excessive, distractible, maddening Pee-wee Herman, as well as, perhaps, the unseemly, morally panicked cautionary tale of his creator, Paul Reubens.

Flashing forward to the digital age, the vision of screen time monstrosity is a little different. If you ask Pixar, maybe we are the obese, babylike space refugees from *Wall-E*. Having exhausted Earth's resources and left it a trash-covered, smog-cloaked wasteland, we blast into space, where we float around on cruise ships, all of our daily lives routed through touch screens on our hover-chairs. Our bodies turned gelatinous by our own lack of embodiment. We became addicted to the speed of technology, only to become near immobile in our reliance upon it. Rather than being wired like Pee-wee Herman, we are wire*less*, floating. But this vision is strained and uncompelling. Its very existence spoils the film. *Wall-E* offers a nostalgic, slapstick masterpiece of digital animation basically right up until these human blobs show up. The wild inventiveness of its dystopian opening, replaced by the madcap obviousness of its screen time satire. They are clichés. Maybe we are clichés too?

But *Wall-E* knows something else about screen time. It knows that screen time can feel good, possibly even, sometimes, under the right circumstances, *be* good. For the film's glorious first half, we watch a little trash robot scurrying around the ruined earth, collecting and compacting waste in a Sisyphean effort to clean the world, to render it new again. He moves with the physical dynamism and charisma of Charlie Chaplin, the grace of Gene Kelly, and his work bears fruit. It's Wall-E who discovers a seedling, the evidence of Earth's renewal. Wall-E is our hero, and when he returns to his metal box every night, he

Figure 1.2. *Wall-E*, 2008

sorts through all the treasured ephemera of humankind he's collected—a cabinet of curiosities to rival the little mermaid—and removes his prized possession: a VHS copy of *Hello, Dolly!* We watch Wall-E watch the film, as close to his monitor as he can be, practicing the gestures of human connection, learning to dance, learning to hold hands. Wall-E's screen time teaches him to be human, and it's only through that that he's able to, literally, save humanity.

Perhaps screens themselves aren't the problem, we worry. Perhaps we are the problem. Our anxieties about ourselves are the anxieties Engelhardt describes about the children of the nineties. If his worry about them is that they would grow up too fast, then perhaps the worry about us is that we haven't grown up at all, unwilling to heed the warnings, incapable of resisting the corporations looking out at us from our multiplying screens, insensible to our complicity in global capitalism. We don't pay enough attention, to our lives, to the natural world around us, to our own children if we have them. If screen time was a moral panic in the nineties, it's a moral failing now. My Screen Time app tells me I pick up my phone, on average, 178 times a day the same way a different app tells me I'm not exercising enough. I try to keep my "pickups" below the dotted line the way I try to not bite my nails or curse in front of my kids. Unhealthy diet, litany of sins, same difference.

In the time since Engelhardt's essay, "screen time" has naturalized into both a piece of received parenting wisdom, a metric for our adult lives, and a creeping vice for us all. But, in the streaming era—and particularly the COVID streaming era—when screens are how some people go to school or go to work or go to the grocery store, nearly every aspect of our day is necessarily mediated by these bright surfaces.

Whether it was ever the real villain in the first place—or a convenient proxy for larger, largely unaddressed familial, social, or institutional failures—screen time is now all the time. What kind of *time* is screen time? Engelhardt is right to point to its organizational pull. Screen time is, first and foremost, a constraint. It's a style of management, a catalogue of mental waste motions, an organization or drawing of boundaries, a way to separate out good time from bad time. It's a way of talking about ourselves. It's a

Figure 1.3. *Mad Men*, "Ladies Room," 2007

way of seeing. We endeavor to control it for kids; an app on our phones endeavors to do the same for us. Creatures made of screens.

The TV is a piece of furniture. Wood-paneled, stately, modern in its midcentury living room. And the kids are staring at it. It's the archetypal image of brainless thrall to the televisual gods. The TV as babysitter, the unseen parents ceding control of their children, the children themselves as zombies—this, too, is screen time. These particular zombies are Sally and Bobby Draper—the children of ad man Don Draper and his wife, Betty. The scene works because of how familiar we all are with the images it stages. A lot of early *Mad Men* is preoccupied with theatrically presenting, and smugly chortling at, 1950s behavior as a litany of now recognizable vices. The smoking, the drinking, the casual sexism, the casual racism, the casual littering—*Mad Men* loved to stage scenes of "normal," white, midcentury life rendered grotesque by the sudden appearance of corrosive customs like these. At worst, these moments were about providing cheap, dramatic-irony-soaked spectacle for contemporary viewers to feel superior about; at best, these moments came to symbolize a creeping rot that contemporary viewers could not themselves escape.

Either way, the origin of American culture's addiction to television is one of those marquee vices. Framed with all the warmth and domestic detail of a Norman Rockwell painting, this image is meant to convey both the ordinariness and menace of that origin. And Betty Draper—archetypal fifties housewife—is *Mad Men*'s embodiment of ordinary menace. Though she may look like Cinderella, Betty Draper always acted like her wicked stepmother. She'd threaten to chop her children's fingers off, destroy their best friendships out of jealousy, do all the worst things that contemporary viewers might associate with the traumatic parenting philosophies of the era: drinking and smoking while pregnant, slapping her kids, conjuring body issues for her young daughter like a malevolent sorceress. And all of these misdeeds were represented by one simple, repeated phrase: "Go watch TV."

Mad Men's obsession with Betty being a bad mom was such a conspicuous part of the show that it rose almost to the level of parody. This character—

sculpted in marble by the extraordinary January Jones—was asked to do so much maniacal, quotidian evil over the course of the series that it began to feel like a bizarre kind of punishment from the writers. She was asked to embody the cultural predisposition to blame mothers for the ostensible sins of technology, to shame housewives for buying and using the products advertised to them. If you google "Betty Draper Go Watch TV" today, you'll find an endless scroll of essays and listicles about her as an iconic Bad Mom: "Betty Draper's 10 Worst Parenting Moments"; "Is Mad Men's Betty Draper TV's Worst Mom"; even a segment on *The Today Show* called "Betty Draper Wins the Bad Parenting Award." (It's worth noting that Don, an equally bad parent, did not receive the same level of think-piece treatment throughout the show's run.) The Draper home's lawless screen time universe made Betty an arch domestic villain, and, vice versa, screen time's real danger was emphasized by its exclusive association with Betty.

Why, at the beginning of the twenty-first century, were Betty Draper's period parenting fails such delectable clickbait? The answer is that, in addition to being playful criticism and flashy content about a beloved television series, these pieces doubled as entries into the popular genre of screen time self-help literature. This is a mode of writing often explicitly targeted at parents of young children, but, crucial to its appeal, is an implicit or explicit acknowledgment of the potential reader's ambivalent relationship to their own (possibly traumatic) screen time history.

In the decades following Engelhardt's initial manifesto, books and articles and television documentaries about kids and screen time proliferated at an unbelievable pace. I became a parent myself in 2015, and I can say that, in the year leading up to the birth of our daughter (the same year as *Mad Men*'s last season), I spent countless hours staring at screens—my laptop, my phone—reading viral articles and listicles about how much screen time it'd be okay for my daughter to have when she finally showed up in our living room.

Engelhardt, of course, was not the first person to manifest anxiety around TV and screens. Fear of television is as old as television itself. Right alongside concerns about the moral content of television programming, commentators and parents have always had concerns about overexposure, from caricatures of bug-eyed kids in the fifties to serious sociological studies about TV's effects on young minds in the sixties to the "Television Free" movement in the seventies. But the idea that parents would rigorously manage their kids' TV consumption, and even that adults themselves would view TV like junk food they needed to apportion and reduce, would evolve over the history of the medium.

As media critic Sonia Livingstone notes, screen time, as we know it, was not initially the central concern of commentators. Initially, our talking heads focused more on immoral influence or physical fitness or even mental health rather than on the raw tally of screen hours. The idea of *time* as an all-encompassing umbrella for all our worst fears about television grew more slowly across the decades. So while Engelhardt's essay did not invent this worry, ironically, it helped to brand an increasing bundle of anxieties with a single, flexible term. And all the condensed vices that he consolidated into the idea of screen time are what twenty-first-century viewers hear every time we hear Betty Draper, or any parent, say "go watch TV."

His essay was thus both a warning about screen time and an advertisement for the concept. The term "screen time" was capacious and nonspecific enough to allow for growth, both in the number of screens and in the age of viewers affected. By 1991, there was already a sense that the anxiety about television needed to be expanded to be more inclusive of other similarly addictive, similarly compromising and compromised screen entertainments. The first home video game consoles, for instance, debuted in the 1970s, but brands like Atari and Nintendo soared in popularity in the latter half of the eighties. Video game arcades housed screen time outside the home, and the terrifying specter of unsupervised, overstimulated teens caused "arcade panics" in malls across America throughout the eighties, as Alexandra Lange narrates. Gaming consoles brought this dangerous new media straight back to the hearth.

Then came the consoles kids could carry with them, and then the consoles that could feel them, register their own physical movements electronically. It was the same physical screen, but it meant something different, possibly more insidious, certainly more intimate. The threats that parents, and everyone, needed to ward against grew ever more powerful and more diffuse.

And, of course, the technology continued apace, introducing second and third and fourth screens into many American homes. Within a few years of Engelhardt's essay, home computing would become widespread, and the emergence of home internet would further open and empower those new screens. Then came the screens for desks and laps and bags and pockets and eyeglasses. The unchecked spread of new screen media in the years following that initial essay, and the increasing public worry about what their spread might signify socially, meant that that essay was continually being written and rewritten, revised and expanded, updated and backdated, targeted at parents and also us all.

Thinking about kids and screens is an easy way to avoid thinking about our own relationship to screens; it's also a good way to do so surreptitiously. Every millennial and a good chunk of Generation X grew up firmly ensconced in this screen time era, surrounded by the culture Engelhardt describes. And so, in the spirit of reassurance, there are plenty of ecumenical books for parents of digital age kids—

"digital natives"—validating any number of approaches to screen time. Some books, of course, offer dire warnings—Nicholas Kardaras, author of *Glow Kids,* calls screens "digital heroin"—but others traffic in a kind of tech-positivity, serving as guides to encouraging healthier relationships between your kids and their screens. Jordan Shapiro, author of *The New Childhood,* writes about gaming with his kids as a new form of connection unavailable to older generations of parents. And Anya Kamenetz, author of *The Art of Screen Time: How Your Family Can Balance Digital Media and Real Life,* even adapts Michael Pollan's advice about healthy eating: "Enjoy screens. Not too much. Mostly with others."

Are screens a form of sustenance, a drug, a contaminant of some other kind? It's not just trade press accounts that try to explain screens metaphorically. Even in scholarly research, studies on screen exposure have always been framed around analogies, often hyperbolic or partially considered. If screen time is a problem, is it more like an addiction or a poison? Is it a cause of disorder or a disorder itself? What existing terms can we use to adequately describe this phenomenon that's so easy to perceive but so hard to quantify? Do we need new terms?

As Kamenatz notes, a lot of these studies begin with the premise that there *is* a problem, that screen time feels so much like a crisis that it simply must be, and proceed with the question of what *kind* of crisis it is. The perception that screens are dangerous, in other words, has real-world implications totally separate from the question of whether or not they *are* dangerous in a verifiable way. Screen time, as we know, just *feels bad* and lots of arguments against it—popular and scholarly—are attempts to reverse engineer the reason for that to be so true.

There are writers who are concerned that excessive screen time can lead to childhood obesity, on the logic that kids who watch too much TV do so at the expense of playing outside. There are writers who suppose that TV is the cause of increased adolescent aggression, even the kind of violent impulses that turn into school shootings. Screen time, we're told, causes poor academic performance; it causes ADHD; it causes distraction; it causes obsessions; it causes anxiety; it causes laziness. Mothers who rely on it are bad. People who indulge it aren't caring well for themselves.

Most of these ills are framed as symptoms of excessive screen exposure rather than any particular screen content. (Screen images don't need to be violent in order to promote violence.) But Engelhardt's initial concerns about screen time were very specific to what was on screen. Sure, "The Primal Screen" was worried about kids turning into vegetables in front of the tube, but it was mostly worried about capitalism, the insidious ways corporations might capture the minds of children. This panicked style of parenting advice that tallies raw hours and sets time limits not only drifts away from an important facet of the original argument, but risks reproducing the very thing Engelhardt was so worked up about in the first place. By focusing so much on time itself—its waste, its value, its expenditure—screen time becomes a concern about productivity. A person who is losing time to screens—perhaps becoming mentally or physically disabled by that interaction—is a bad worker and a bad consumer. They themselves become a waste. Screen time is theft.

The common thread among all of these potential dangers is that while all of them may *seem* as though they'd be caused by excessive screen time, there's very little hard evidence that they are. It's not that there are mountains of evidence pointing to the positive impact of screen time, but there simply isn't a mountain of evidence for the opposite. Indeed, despite decades of research, the current state of the field can be pretty well summarized by the opening line of *Television in the Lives of Our Children*, a study published in 1961: "No informed person can simply say that television is bad or that it is good for children."

For Kamenetz, reading through this same vast popular and scholarly literature, the phenomenon with the most verifiable linkage to screens isn't anything so flashy as violence or hyperactivity. It's actually sleep that emerges as the biggest and most statistically significant concern. If the connection between screen time and all of those social and individual ills is hard to pin down, the question of screens' influence on sleep is refreshingly direct.

But to see it, you need to stop considering screens as metaphors and see them as material objects. Blue-spectrum light, like the kind produced by most screens, messes with the production of melatonin, and a deficit of melatonin can make it hard to fall asleep. Simple as that. Poor sleep, in turn, can lead to a cornucopia of problems, many of which are often ascribed directly to screen use.

But poor sleep has a variety of causes. Lisa Guernsey, for instance, cites a study that shows just as many kids reporting trouble falling asleep due to reading *Harry Potter* books as watching screens. All the same, sleep specialists are adamant that, regardless of screen diet overall, simply refraining from screens before bed is enough to dramatically improve outcomes on any number of fronts. And this seems to be good advice, for what it's worth, for kids of all ages.

For a variety of reasons, it's very difficult to design experiments or enlist participants to scientifically test the effects of screen time. And even when negative outcomes can be identified, studies tend to present a blurry picture of causation. The popular literature on screen time, then, rather than offering firm proscriptions and proven causes, tends to be mostly interpretive. There's simply not that much evidence that screens, in and of themselves, are harmful to anybody. Betty Draper's living room is a dangerous place, but it's not necessarily the TV set's fault. So while many of these books differ from Engelhardt's cultural provocation by rooting their ar-

Figure 1.4. *Mad Men*, "Waterloo," 2014

guments in studies and statistics, most of them are just echoing what Engelhardt was saying in the first place: growing up is hard to do.

So Neil Armstrong lands on the moon, and Peggy Olson has to wake up and do her job the next day, which is to pitch a television advertisement to a Midwestern fast food chain called Burger Chef. She's already expressed her anxiety about it. "I have to talk to people who just touched the face of God about hamburgers," she laments. But, rather than avoid the subject, she leans into it. "I don't know what was more miraculous," she begins, "the technological achievement that put our species in a new perspective or the fact that all of us were doing the same thing at the same time. Sitting in this room, we can still feel the pleasure of that connection." This pitch isn't about the moon, it's about television. But it's an ambivalent vision of TV's cultural role. On one hand, the TV allowed the world to unite in awe and reverence, but on the other hand, the TV is a distraction from real life. "I don't need to charge you for a research report that tells you that most television sets are not more than six feet away from the dinner table," she continues. "And that dinner table is your battlefield and your prize. This is the home your customers really live in. This is your dinner table."

Screen time makes us feel good, Peggy acknowledges to the executives at Burger Chef, but maybe it shouldn't. So she acknowledges the connection TV provides, but she tries to envision a way to achieve it without the TV as medium. "What if there was another table," she imagines, "where everybody gets what they want when they want it. It's bright and clean, and there's no laundry, no telephone, and no TV. And we can have the connection that we're hungry for."

I mentioned that screen time began as one of *Mad Men*'s favorite vices to ridicule. Over the course of the series, however, it became more and more clear that screen time was actually one of the show's primary subjects. If you want to imagine what creatures made of screens look like, what better demographic is there than the people who make television commercials. And Peggy's monologue is perhaps its greatest statement. In the space of this pitch, we find television as a space of connection, the moment when the world gathers together to look beyond itself. The Moon landing is a genuine, collective revelation, enabled by the technology of television—screen time as transcendence. But we also have television as fraud. The TV is the *substitute* for connection, the fake hearth, a medium for communication that shouldn't need a medium. Peggy's speaking to the Burger Chef executives, but it's visually framed as though she's speaking to us, watching *Mad Men* at home. "This is the home your customers really live in," she says, "the TV's always on." And the utopian pitch that she makes, to those of us watching her inside the screen and out, is for a place without it.

Of course, the irony is that this itself is an advertisement. She wants to help Burger Chef's customers to flee the prison of the television for the fluorescent lights of a fast food chain. And this romantic vision of a TV-less place is designed to be propagated *on* television. The scene ends as Peggy begins to describe the TV commercial shot for shot. TV tells us what our desires are and helps us to fill them. We see the first storyboard—framed with the same rounded edges as Don and Betty's TV at home—from the perspective of Burger Chef, the viewer, the customer. *The TV's always on.*

For nearly two years, though, in our house at least, it wasn't. When our oldest daughter, Maeve, was born, our main concern was her eyeballs. At the time, the American Academy of Pediatrics' official recommendation was to hold off on screens of any kind until the age of two. This turned out to be a pretty controversial position when the AAP launched it in 1999. As Lisa Guernsey writes, the organization received a tremendous amount of pushback from families, and, what's worse, it put them in the position of having to vigorously defend a recommendation that was more best guess than hard science. The two-year prohibition, as the AAP would readily admit, was not the result of any real research suggesting particularly terrible outcomes for toddlers exposed to screens. Rather, the AAP made the rule because they could find no research that said it was particularly *beneficial* to show screens to tiny kids. Their hope was that the ruling would have the effect of encouraging parents to lovingly and attentively care for their children—things we know are good—rather than taking the Betty Draper approach. It was parenting advice disguised as medical advice. In 2016, the AAP relaxed their prohibition.

When Maeve was born in 2015, we had planned on being loving and attentive regardless, but we decided to try the AAP policy anyway. We were not particularly worried about violent images, a sedentary lifestyle, or hyperactivity. Though we desperately wanted her to sleep, we were not yet worried about how a screen might affect that. We were, however—spurred on by a number of articles we read on our phones seconds before trying to fall asleep—worried about whether early exposure to screens might damage her tiny little rods and cones. Everything about a baby is fragile, even for a nine-pound, two-foot galoot like the one we had. I was inherently suspicious of the moral panic arguments about screen time. But I understand them better now than I did before becoming a parent. Specifically, I understand the impulse to interpret every stimulus as a potential danger, regardless of whether they are or not. Kids are small, we ought to protect them. That seems pretty basic. When even a galloping toddler gets a new tooth or a sniffle, their fever can spike into the hundreds. It wasn't hard, looking through a brand-new set of parenting goggles, to imagine the TV's malevolent rays cooking my daughter's young mind, sautéing her baby blues.

So we decided to give it a try, our own Screen Time Zero policy. For most of those first two years, Maeve must have thought about our TV the way that the prehistoric human ancestors thought about the black monolith in Stanley Kubrick's *2001: A Space Odyssey*. Just a giant, dark, possibly nefarious void. I say this with all the cringe in the entire world, but it likely just seemed like a big, black mirror to her. And, weirdest of all, it wasn't that hard. Mel and I watched TV after she went to bed at night, and, when the weather was bad, we tuned into Steve Templeton, our trusted meteorologist, but otherwise, it stayed off. And Maeve had books and toys and a couple cats and grandparents and all sorts of stuff to stay occupied. Here was this little person who showed up, and she didn't ask us for screens. She hadn't even heard of them.

But it couldn't last forever—nor would we have wanted it to—and, in fact, we broke a bit early. The first time was a necessity. Three months before her second birthday, we were preparing for a road trip. We live out in the middle of the country in St. Louis—where a good local meteorologist like Steve Templeton is important—and we were planning to drive to a rented cabin in Kentucky to meet up with Mel's family, who would be traveling to meet us from their home in Philadelphia. For this trip, we knew that we would need something to occupy her beyond blocks and coloring books and the majesty of the American landscape. So we gave her a tiny iPad that Mel had gotten as a free gift when she replaced her phone. The iPad had three things loaded onto it: one season of *Wonder Pets*, one season of *Daniel Tiger's Neighborhood*, and *Moana*.

And it worked like a charm. Maeve delighted at this screen that suddenly came alive in her hands. She marveled at the sight of a little hamster and turtle and duckling flying around the world to save baby animals; she learned Daniel Tiger's little songs about sharing and being patient and pooping on the potty; she got lost in the sparkling seas of Disney Animation Studios. She was distracted from the arduous hours of driving, but I don't think it'd be right to describe *her* as distracted during that first real stretch of screen time. If anything, she was more attentive. There was so much to see, so much to discover, so much to pay attention to.

That sense of possibility, obviously, doesn't last forever. Maybe you think the train is barreling off the screen at you the first time you see a motion picture, but that doesn't happen again. We gave Maeve that little screen as a utility. "Go watch TV," we told her. But if we think of screens merely as utilities, misused or corrupted tools, we miss something essential about them.

That was just an iPad, though. The TV itself—the monolith—turned into a TV when the Philadelphia Eagles won the Super Bowl a few months later. We were back in St. Louis, and Mel's hometown team was, improbably, playing for a championship. This was a big moment for Philadelphia, and a big moment for Mel to be so far away from it. Because the game was happening before Maeve's bedtime, we decided to present the event as what the Super Bowl is to anyone who watches it: a ritual. We set Maeve up with a little table, surrounded by snacks and puzzles, we taught her when to cheer, Mel specifically taught her how to boo like a Philadelphian, and we all watched, conspicuously, performatively, together.

Maeve was obviously lost by the game itself. What she watched most was her mother. Mel herself watched intently, leaping from the couch spontaneously, sometimes hooting in victory, sometimes muttering recrimination. Sometimes, she and Maeve would lock eyes, Mel with the knowledge that this act of co-viewing was something she'd remember for a long time, Maeve with a look of amused curiosity at the whole spectacle. Sometimes, Mel would pull out her phone, a second screen, to text and celebrate with her brother and her best bud Kevin, all the way back in Philly. One thing Maeve knew about us both, even at the age of two, was that we were both far from where we are from. She could perceive, I think, the difference in our faces when we'd reach the endpoint of our Pennsylvania road trips later on, and I bet she saw something of that face in Mel, in the winter, in our St. Louis apartment. When Maeve watched her mom watch the Super Bowl, I don't think she saw a zombie. She saw somebody who, for a couple of hours, felt back at home. *The fact that all of us were doing the same thing at the same time. Sitting in this room, we can still feel the pleasure of that connection.* This is screen time, too.

2

THE SCREEN TIME ERA

When I was little, I learned to tell time through television.

Whenever I needed to count things down—getting ready to withstand a long trip or frenziedly waiting for a friend to arrive for a playdate—my parents would give me an estimate based on the unit of time that I knew more intimately than any other: the length of a single episode of *DuckTales*. I would watch *DuckTales*, Disney's 1987 cartoon series, in our basement, on a large, wood-paneled RCA Colortrak TV set that could swivel. I loved *DuckTales*, and, like

Figure 2.1. *Bluey*, "Takeaway," 2018

anything you love, I developed a sense of what to expect from it, what I wanted, what gave me pleasure. I can still remember the satisfying liquid crunch of Scrooge McDuck nose-diving into his vault filled with gold coins. I still hear the theme song—*awoo-ooh*—rattling around in my head. I knew exactly, and still know very instinctively now, how long an episode lasts.

So when I'd wonder out loud how long something was going to take, my parents could tell me that, "it's another two DuckTales until we get there" or "he'll be here in four DuckTales." Eventually, I'd just ask, "How many DuckTales is it?" A DuckTale, to me, was a meaningful, and deeply familiar, duration. This is a different kind of time than the one Maeve learned watching the Super Bowl with us—not the "now" but the "when."

This bodily awareness of televisual time, that we learn *through* the television, isn't unusual. *Creatures made of screens.* Even cartoon dogs have it. There's an episode of *Bluey*—one of the only shows Maeve and her younger sister, Phoebe, will consent to watch together—where sisters Bluey and Bingo are waiting to pick up takeout at a Chinese restaurant with their dad, Bandit. Bandit tells them it'll be five minutes, and Bingo, squirming, asks her dad how long five minutes is. He thinks for a minute, and then replies, "It's one episode of *Chutney Chimp.*"

Maeve and Phoebe, in turn, often tell time by way of *Bluey*. They know that there's a difference, for instance, between how much time they'll be cuddled on the couch if we're watching an episode of *Bluey* (seven minutes) versus an episode of *Spirit: Riding Free* (around twenty-two minutes). They budget their own time, budget what they're asking for, based on those time scales. They'll ask for *Bluey* if they know that they're trying to squeeze in a quick episode after dinner or before we have to be somewhere; they'll ask for *Spirit*—which, unlike *Bluey*, is both longer and serialized across episodes—if there's a more open-ended expanse of time ahead. Before we could read a clock, before we could set a timer, before we could reliably sort out the difference between minutes and hours, TV provided a template in time slots. And, then, DuckTale after DuckTale, Bluey after Bluey we learned—are learning—what time is, what it's for, and how much a unit of it can matter.

Not every screen is a television. But TVs have always been, quite literally, both form and content. So it was both TV shows and TV sets that inaugurated the relationship—between viewers, screens, and space—we all now occupy. Every new personal screen has evolved in the shadow of those wood-paneled originals. As Lynn Spigel notes in her sweeping history of televisions and families in the postwar US, Admiral television sets were built into the design of some of the very first suburban homes in Levittown, New Jersey, in the fifties—they were in the walls, part of the actual architecture of suburbia. And they were as much a part of the idea of suburbia as they were a part of its brick and mortar existence. The TV was a hearth, a communal space; it was also a sign of modernity and progress. As it changed, moved around the house, proliferated, it always retained that dual physical and psychological significance.

Indeed, regarding the television set as a new "hearth" has been a key rhetorical move for partisans on both sides of the screen time wars since the nineties. (The critic Cecilia Tichi christened the TV the "electronic hearth" the same year Engelhardt published "The Primal Screen.") On one hand, maybe the TV set is a cheap, commodified substitute, replacing the nourishing warmth of the family fire with the dull, thrumming heat of an electric tube; on the other hand, maybe the TV set is better than some symbolic hearth, because it engages the family, gives them something to do together, something more materially communal than sitting around a hole in the wall. Televisions are a load-bearing piece of furniture for the postwar American psyche, and that's only become more true as the idea of "television" has become more diffuse.

We've talked already about what people are worried about when they worry about screen time. They're worried about control, about contamination, about kids growing up the wrong way, or just a different way than they did. They're worried about physical fitness and mental health and eyesight and violence and sleep and consumerism. These are all things that screens and screen media are imagined to degrade or damage or devilishly encourage if we spend too much time with them.

But what about the screens themselves? How do screens spend *their* time? We try to protect our kids and ourselves from wasting too much time on screens, but it's also true that screens have irreparably altered the way we understand what time even is.

As media theorist Shane Denson puts it, screen time, as a concept, "links the power of image, as an explicitly temporal force with . . . the formation of the 'subject.'" Screen time, in other words, is a form of education or coming of age, a way to understand how media exposure, for good and ill, helps us become who we are. The critic Amy Holdsworth even argues that the experience of life itself can become televisual in the presence of all these screens: "looping, repetitive, banal, catastrophic, messy, incomplete." Because screen time—formative, but messy—is so closely linked with growing up, the solutions our culture has posed to the "problem" of screen time have thus inevitably tilted toward regulation and constraint.

But, funnily enough, television itself is also a medium of regulation and constraint. So if we want to think about the impact screen time has on our own growth and development, we need to think a little about screen time's own coming of age, too, starting with television. Its various genres are often defined by their temporal boundaries: the half-hour sitcom, the hour-long drama, the limited miniseries, the live broadcast. They're defined by the hour they're designed to air: daytime, prime time, late night. More dramatically than even the theater or the Victorian serial, and just as much as radio drama, the most instantly recognizable modes of TV, even today, were shaped in their infancy by the simple question of how much time is available to show them, when, and over how long a period.

And that's only thinking about questions of length and duration. These forms also evolved historically in relation to time slots, commercial breaks, or even seasons of the year. Aristotle certainly influenced the act structure of the television episode, but so did Chevy and Coca-Cola. The origin stories of TV's temporalities belong to the old-school network era, but they remain, like vestigial tails, in what Aymar Jean Christian calls the contemporary "networked era."

Here's my point: the history of television is a history of how those constraints became generative, rather than limiting. Television dilates the space given to it; it makes time malleable. Its time frames speed us up and slow us down, expand our time or help us lose it. And so, just like time slots or commercial breaks, when "screen time" was codified as a popular concept, it too became a formative constraint for television.

Screen time is a mode of organizing time, a way of policing it, a way of assigning value to it. As a valuable possession, it is currency, and it's a currency that can be misspent, taxed, squandered, invested smartly or foolishly. It is a life measured out in DuckTales. And it is naturally—or unnaturally—fast at its excessive boundaries. Too fast. But inasmuch as screen time is a kind of time, it's also part of historical time. People have been worried about TV, about screens, since these things have existed, but—as I've said—the modern form of these debates we date to the nineties. To lend a name to something is to lend it a coherence it may have lacked before, and this anxiety, this possibility, this *time* was given a name in 1991, and it still answers to that name today: *the screen time era.*

But, as we know, screen time is not just a kind of time; it's also a feeling. The viewer of the screen time era is acutely and increasingly aware of *how much* they're watching, but they're also aware of how that quantity of screen time makes them feel. This is a viewer whose guilt—pleasurable or otherwise—about their own spectatorship is an active part of their point of view. And so, this viewer increases their connoisseurship, their fastidious supervision, about the kinds of programs in which they will invest their screen time. They are in control, or, at least, they value the *feeling* of control. But this is also a viewer who is potentially out of control: binging, scrolling, hate watching. The viewer of the screen time era, regardless, is incredibly self-conscious about their viewership, its investment of time, in an anxious way.

And the programs they watch have evolved to mirror that self-consciousness, and that anxiety, too.

So how has the screen time era shaped, not only the way people watch screens, but what people have put on screens for them to watch? What forms are unique to the screen time ecosystem? What kind of things have artists, networks, tech companies thought to create to cater to a viewer or consumer who lives in this atmosphere of anxiety about and personal regulation of screens? All your favorite hits from the nineties to today: prestige TV serials, *House Hunters*, TiVo, Nintendo Wii, CGI preschool cartoons, *The Real World*, Microsoft Word and the Notes app, Vine, YouTube videos of UFOs, iMessage, Wikipedia, the Twitter scroll, Google Maps, *Pokémon Go*, Peloton, Twitch, Sky Guide, and every other GPS app that tells you where you are in the world through the imagined animation of a screen. These are the forms of the screen time era.

* * *

There's a fictional TV show on the TV show *Twin Peaks*. It's a soap opera called *Invitation to Love*, and, at various times, characters on *Twin Peaks* watch the show or have it on in the background of their houses. Like *Twin Peaks*, *Invitation to Love* is a bit of a sensation. We don't know a great amount of detail about *Invitation to Love* aside from the fact that it seems tailor-made to replicate the style and substance of the typical daytime soap of the 1980s: soft focus, histrionic dialogue, occasional melodramatic swells of passionate sex or equally passionate violence. When we see it onscreen, it's always doing something. (The critic Forrest Wickman argues that *Invitation to Love* is the "shadow self" of *Twin Peaks*.) Televisions are pieces of furniture in the homes of the ordinary and strange people of Twin Peaks, Washington, and *Invitation to Love* is both evidence of their ordinariness and rhythmic counterpoint to their strangeness.

Talking in 1991 about the decision to build space for this show-within-a-show, Mark Frost, the TV veteran who co-created *Twin Peaks* with filmmaker David Lynch, said, "I think that watching television is a big part of people's lives in this country, and you very rarely see that treated in television." *Twin Peaks* is a show that's very aware of itself as a television show, and, as Frost says, it's aware of its characters themselves as television viewers. The show never takes this particular part of modern subjectivity for granted.

It's also a series that foregrounds the way that television—certainly its midcentury golden era, filled alternately with scenes of domestic tranquility and cops and robbers—influences our assumptions about the lives we live: what types of things do we expect from a "quiet, small town," what is a cop *like*, what is a murderer *like*, what are the beats of a budding romance? TV shows, like movies or novels or plays, construct a world that is the right shape for the stories they tell. Those worlds have narrative rules and cosmic laws and social norms, and part of the way TV series invest us in their stories is simply by teaching us about those rules and laws and norms.

Eventually, if enough televisual worlds are built on the same model, then those worlds begin to influence the way we, as viewers, see our own worlds. TV shows teach us how to watch them, but that pedagogy inevitability extends outside of our viewing. It's a cliché, but the media we consume has a big impact on the way we see the world. *Twin Peaks* is, in some tangential way, *about* how that truth plays out for its characters.

Twin Peaks, an objectively weird show, managed to become a breakaway hit on broadcast TV largely because it improvised, unnervingly, self-consciously, with the genre conventions of the daytime soap and the cop show. Yes, *Twin Peaks* performed the paradigm-shifting, avant-garde televisual experiment its cult fans now praise; it also was an ingeniously but pretty conventionally structured TV serial, a prime time soap. Its shape was recognizable despite the uncanniness of its contents. And while that shape itself got stranger over the course of the series, its innovations were smuggled in initially through a form that was ultimately familiar to viewers.

The show's occasionally violent, frequently surreal deviations from the norm are deviations from a norm culturally defined, in part, by television and its deeply familiar genres. If the people of Twin Peaks have lost control of their lives, they've lost the specific type of control that television series—soaps, serials, procedurals—promised. What *Twin Peaks* fractures is both the form and the utopian vision of classic television. The TV fantasy becomes a TV nightmare.

It's not a coincidence, then, that, in solving the central mystery of the show—the disappearance and murder of Laura Palmer—Frost and Lynch choose to frame the horror of their supernatural villain by way of scenes of terrifying and terrified spectatorship. Perhaps the most iconic of these scenes occurs in the second episode of the second season. Maddy (played by Sheryl Lee, who also plays the murdered Laura Palmer) crouches down in the corner of a living room and has a vision of Bob, the malevolent, long-haired denim-on-denim demon that's responsible for Laura's murder as well as many other calamities in the area. Bob's a physical being, but also a manifestation of evil—hence the abomination of his Canadian tuxedo—so, when he's onscreen, the narrative reality of the series warps around him.

The scene is two long shots from Maddy's point of view, as Bob appears in the background, makes eye contact with Maddy, and proceeds to slowly, menacingly climb over the couch in the center of the frame toward her, eventually blocking out the camera's lens with his face. It's Maddy's perspective—there's one quick eyeline match early to make sure we know that spatially—but it's impossible not to also think of the scene as playing out from the perspective of a television set.

Maddy curls up behind a chair in the living room, which we can see, out of focus, in the foreground.

But she's also sitting right where a television should be, separated from the couch by a coffee table, visible from the dining room. The shot's almost perfectly symmetrical, with the sofa in the center of the frame—we as viewers are positioned with Maddy precisely in the space of the hearth, where manufacturers and designers had forever recommended the placement of the TV itself. And that's not to mention the fact that Bob's unbroken eye contact with Maddy immediately becomes unbroken eye contact with us. Bob's creeping approach is thus a kind of Ouroboros moment for the series, the television eating its own tail. Maddy is terrified, we are terrified. Bob crawls toward us from what appears to be a mirror image of our own domestic spaces, two living rooms connected by their televisual navels. The fourth wall, which Bob is both breaking and not breaking, is the screen itself, now a window or a door.

We are the TV. He's looking at us.

The year that Tom Engelhardt published his screen time essay is also the year that *Twin Peaks* went off the air in the US. I have to wonder if Engelhardt watched it. In retrospect, it's fascinating to consider Engelhardt and Frost/Lynch as parallel theorists, the sociologist and the surrealists, of the way our time with screens shapes us.

So then maybe it's appropriate to consider how both Engelhardt and *Twin Peaks*, together, influenced a new age of television. That age goes by many names: the "golden age" of television, the "second

Figure 2.2. *Twin Peaks*, "Coma," 1990

golden age" of television, the "silver age" of television, the age of "quality" TV, the age of "prestige" TV, the age of "complex" TV. But regardless of what we call it, this historical period of TV production that's been named and renamed, analyzed and obsessed over, that's altered the very structure of television itself as a medium, is squarely a part of the screen time era. And it shows.

Whether or not you mark *Twin Peaks* specifically as the genesis moment, the hour-long dramas of the late eighties and early nineties represent a prelude and beginning that would come to its recognizable shape a few years later with *The Sopranos* and extend ever outward to include *Mad Men, Breaking Bad,*

Game of Thrones, *Succession*, and the other premium cable and streaming series that have done so much to define the conversation around twenty-first century television.

Perhaps the easiest way to situate this era is to start with premium cable's aggressive interest in producing original dramatic programming at the end of the century. There's a somewhat famous anecdote, relayed by Alan Sepinwall in his book *The Revolution Was Televised*, that HBO's executives would recruit network TV writers and producers in the late nineties by simply asking them what they weren't allowed to do on broadcast TV. The most obvious answers were sex and violence. And HBO's early, paradigm-shifting experiments—David Chase's *The Sopranos*, David Simon's *The Wire*, David Milch's *Deadwood*—lustily embraced this new freedom.

The emergence of the "antihero" series—which often comes to stand in for a lot of the other innovations of this era—is precisely the type of TV you'd design if you were trying to wedge in as much sex and violence as you could. But the less obvious answer, and the most important one for all of these shows and the ones that would follow—for the way all these shows would relate to our feelings about our screen time—was seriality.

Jason Mittell—who coined the term "complex TV"—defines these shows by what their narrative structures ask of viewers. Complex TV, in his description, isn't just about telling an ongoing story week by week, it's about creating a narrative ecosystem in which the viewer needs to be an active part. TV like this works cumulatively. Each episode pushes the narrative forward, even and especially when it's not entirely clear how. It's on the viewer to remember, to notice, to pick up on the references—visual and otherwise—the show's writers and performers make to previous plot elements or to foreshadowed eventualities. As they say on *The Wire*, "all the pieces matter." And they matter, in part, because you, the viewer, have to be their keeper and closest watcher.

The episodic narrative approach of the familiar sitcoms and procedurals of the fifties through the eighties—the repetitiveness against which these newer series defined themselves—purposefully absolves the viewer of that kind of investment. These shows repeat, they remind, they establish firm formats, and they meet expectations. They never, really, need to end, so they don't even burden the viewer with that sense of apocalyptic foreboding. They thrill or provoke or otherwise entertain, but the responsibility of their success is on the show itself, episode by self-contained episode.

Mittell's "complex" series, on the other hand, need *us* in order to work. They do not repeat or remind or hold to established formats. They are infused with what Frank Kermode called "the sense of an ending." Even their moral arguments—the status of their antiheroes—require us to acknowledge and interrogate our own pleasurable complicity in our favorite characters' crimes. These shows accumulate their narrative momentum on us, on our bodies in time. And, as with our bodies, we know they are lurching, fast or slow, toward their end. All that sex and violence is only heading one place.

So cable networks, seeking to "revolutionize" TV by producing adult programs niche marketed—or, "narrowcast"—to their most prized demographics, chose to adapt the soap opera to prime time in this way. To avoid getting bogged down in value-laden language—like "prestige" or "quality" or even "complexity"—we'll call them prime time serials, the phrase used by critics Michael Z. Newman and Elana Levine. Shows in this form are so common now as to be ubiquitous, but when Agent Dale Cooper arrived in Twin Peaks to investigate the disappearance of Laura Palmer in 1990, the form was still in its first decade. *Twin Peaks* is often lumped together with a set of deliberately, provocatively pathbreaking nineties serials, including *NYPD Blue, ER, The X-Files, My So-Called Life,* and *Buffy the Vampire Slayer,* but, in truth, the changes these high-profile series announced had been in the works since the mid-eighties. Shows like *LA Law, Hill Street Blues,* and *St. Elsewhere* defiantly decoupled their genres—legal drama, cop show, medical drama, respectively—from their traditional, episodic, case-of-the-week structures. (These shows were also actively decoupling from more straightforwardly soapy prime time series like *Dynasty* and *Dallas*.) But what makes these shows different from, say, *Invitation to Love*?

The answer is yet another currency: taste. *Twin Peaks*' narrative innovations might not have been brand-new when it debuted, but they were packaged in a way that was certainly read as novel. *Twin Peaks* was an early-nineties pivot point on the road to what Newman and Levine call the "legitimation" of television, its cultural acceptance as a possible medium for serious art after decades of being described, famously, as a "vast wasteland." *Twin Peaks* was seen by many as fundamentally different amid its peers. But *Twin Peaks* was different in a way that foretold more like it. Newman and Levine call it a cultural "harbinger" that television was on the way to being perceived as becoming, simply, better.

The true turn toward the legitimation of the new TV age was not, in fact, about form or content alone, but about the public perception of artistic superiority. This is where the "prestige" tag comes in. The prime time serials occupied a place at the top of a taste hierarchy that their networks helped to build. HBO's tagline during the mid-aughts heyday of these shows was "It's not TV. It's HBO." The pretentiousness, as well as the playful medium confusion, of that advertising slogan revealed the real game: to make something that viewers would perceive to be qualitatively superior to, even materially different from, television itself.

I don't think it's a coincidence that killer Bob climbs out of his TV screen into your TV screen right before Engelhardt's essay coined "screen time" and let loose the decades-long moral panic in which we find ourselves embroiled. I don't think it's a coincidence that this genre of television emerges historically at the same time that a discourse premised upon the idea of TV as a poisonous environmental toxin gets its name. The sociologist and the surrealists at work.

At every level, the prime time serial lives in counterpoint to screen time, acknowledging and subverting the worst fears of those who think of television as a corrosive cultural force. Here is a genre defined by its quality, its prestige, its complexity, and its interpretive ambiguity. And it needs you to actively, not passively, help construct it: your activity, your intelligence, your engagement. No zombies allowed. Here is TV—producers, writers, actors, and critics even—defending itself.

So, in that sense, the prime time serial is designed to assuage the worries of the worried, particularly those worried about the wise investment—and possible corruption—of their time. What if there was such a thing as TV that was good for you in the way that literature, visual art, or even classical music are good for you? But the prime time serial is also a pure product of the screen time era because of its conspicuous indulgence in some of the worst, most villainized impulses of the time. For as radiant its aura of prestige, the prime time serial is also a precision vehicle for the delivery of boobs and butts and bloody, bloody violence. It's TV that seems like it would be bad for you marketed and valued as media that's good for you.

There's no better example of how this works than what the TV critic and scholar Myles McNutt calls "sexposition." Writing about HBO's massive hit *Game of Thrones*, McNutt noticed that long faux-Shakespearean soliloquys, full of byzantine narrative detail and lurid grasps toward poetic transcendence, were often delivered by characters who—surprise!—find themselves onscreen during graphic sex scenes. Sexposition scenes—which certainly precede *Game of Thrones* itself in the genre—are the convergence point of these series' form and content, the narrative complexity and the titillating sexuality mutually justifying each other. The prime time serial indulges both the virtuous resolutions and the taboo urges of the screen time era viewer.

The most significant way in which the genre is peak screen time is in its obsession with, even fetishization of, attention. What if there was a type of TV that asked for, and rewarded, the type of excessive commitment of time you were already flushing down the drain? What if that TV had no commercials, and what if it engaged your imagination? What if, as Jenny Odell suggests, attention itself was a valuable currency, the product these shows advertised? What if we return to the fifties to learn that the cure for screen time is more, and different, screen time?

Engelhardt wrote that "maturity means the ability to discern." To watch a prime time serial the way it's meant to be watched is to practice the kind of discernment Engelhardt describes here, though I suspect he'd be loath to describe it as such. It's not just about following complicated plots or remembering minor character names or picking up Easter eggs in throwaway moments. TV series of this kind presuppose an attentive viewer, a discerning viewer. More than that, they invite a viewer who *self-identifies* as attentive or discerning. They present to us the opportunity to wonder about what it means to watch *The Sopranos* or *Mad Men* and identify with these bad men, to think about our moral status in our desires. We may not be revolutionary in our insights when we consider that *The Sopranos* is about the collapse of American empire and that *Mad Men* is about the toxic legacy of fifties consumer culture, but we are having insights all the same. These are all aspects of these series that require our interpretation, our engagement, our attention and discernment to reveal.

It's also true that this same quality of attention is requested regardless of the quality of the text. In other words, while lots of these series do interesting and thoughtful work at the level of narrative, they don't all do it equally. The genre demands that we pay the same kind of attention to all of them, though. Our discernment authorizes every show no matter how artfully that show rewards it.

The viewer of the prime time serial is not passive—or, more precisely, the viewer of the prime time serial is reassured, flattered, that they're not passive. To watch the prime time serial is to be *doing something*, or at least to *feel* like you're doing something. You feel like you're spending your time in a way that's under your consideration, if not fully your control.

And the clearest way to see that is the genre of writing that rose up to meet these shows: the TV recap. Recaps were designed initially to recapitulate plot points for fans. But, over time, they evolved as episodic reviews, even essays, that paid intense, granular attention to each step of these serialized narratives. (Making the screen time era resonance even more prominent, the critic Matt Zoller Seitz calls them "overnight reviews.")

But while eventually getting co-opted into the prestige ecosystem, the recap form emerged out of fan communities that very explicitly did not consider themselves highbrow. The generic model was developed in the twentieth century through fan magazines like *Soap Opera Digest*. The recaps there served the practical purpose of helping viewers who missed episodes catch up on quickly evolving plots. They were born out of a recognition about precisely the type of "complexity" Mittell ascribes to the prestige series of the twenty-first century: your enjoyment as a viewer depended upon your having caught up.

The mothership of this form in the screen time era was the website Television Without Pity, which posted long recaps of TV episodes mixing plot synopsis with critical evaluation and interpretation. But the most popular forums on TWoP, especially in its early days, were for series who carried their debt to soaps with less self-conscious embarrassment and defensiveness than the series on premium cable. The form came to life in recaps of *Dawson's Creek* and *Buffy*, rather than *Deadwood* and *The Sopranos*.

But, as Newman and Levine point out, the "legitimation" of TV in this period was a gendered process. The disavowal of the soap opera as an influence, for instance—a phenomenon that's been chronicled by feminist media scholars like Tania Modleski and Kristen Warner—demonstrates how intensely our traditional taste hierarchies have understood prestige as the opposite of womanly genres, and arguably womanliness itself. As Sarah Blackwood and Sarah Mesle put it, "taste is just another word for internalized misogyny." Indeed, the type of prestige these shows seek is one rooted in long-held, romantic visions of male genius and creative power.

Despite the crucial importance of female executives, writers, and actors to all of these shows, the idea of a cadre of irascible, white, male geniuses—the Davids of HBO—inventing a new form by making shows about irascible, white, male geniuses has proven, over and again, irresistible to the writers looking to tell the story of television in the twenty-first century. *The New Yorker*'s Emily Nussbaum has even pointed out how these narratives rely on the erasure of *Sex and the City*, a show that brought antiheroes and complex seriality to the sitcom a full year before *The Sopranos* did that for the soap. Recappers, paying such granular, extended attention to these series, had the space and time to focus on the collaborative process of making contemporary television, but the recap was also a small part of a growing TV criticism ecosystem that helped cement the idea of the showrunner or creator as primary author—even *auteur*—of a TV series.

So the TV recap provided the critical architecture for such a narrative to take off, as well as essential spaces for pushback against it. Less focused on broad evaluation or even recommendation, and more on the performance and sharing of attention, TV recaps are the homework of this type of screen time. If this TV is so good for you, *show me.* While those early soap recaps served the practical purpose of catching viewers up on what they missed, the TV recaps of the screen time era were works of analysis meant, largely, for viewers who'd already seen the same episode. Like a lot of viewers, I have my favorite critics I turn to after every episode airs. IndieWire even used to publish recommendation guides for the best recaps of buzzy shows each week so viewers, now readers, could make the rounds to see how their favorite critics saw what they just saw.

This genre of writing was thus leveraged toward solving another of the bogeymen of the screen time era: isolation. The recap, and its comments section, was the place where you came together with a community of other viewers. And that community, for what it's worth, was one that was a lot less dudely than the list of creators of these early twenty-first-century series. If the creative culture of HBO and the other cable networks that followed was dominated by men, the critical culture that rose to greet it avowedly was not. A lot of the energy surrounding episodic criticism came from the way the genre became an ideal space for an emerging, popular feminist critical sphere that would draw attention to the dynamics of race, gender, and sexuality undergirding these shows. The canonical early geniuses of prestige TV were white men like the Davids, but the people who processed their work in public weren't always.

All the same, the recap also came to be about the kind of legitimation Newman and Levine describe. If recaps of *Dawson's Creek* were never primarily about making *Dawson's Creek* seem aesthetically significant, the attention they represented nevertheless came to be absorbed into a high-culture economy. This kind of attention, especially when paid by critics at high-end glossy magazines and prestigious outlets, proves its value. It serves as an implicit argument for the artistry of these series. Why would serious people spend this much energy dissecting and analyzing and debating these series if the series weren't serious themselves? And that's where all of this lands. Can a TV genre transform this medium from a repository of cultural detritus into a shining beacon of cultural capital? Screen time is a currency. What's it worth?

* * *

When television shows center the particular subjectivity of the television viewer, they become really interesting to people—recappers, critics—who are hyper-aware of themselves as television viewers. Another way to put this is that the more shows become about spectatorship, the more they are fit for the recap, the genre of writing most at home in and most legitimating of the screen time era. They make criticism seem necessary, glamorous, even meaningful to viewers and creators alike. These shows aren't just good prompts for TV criticism; they *are* TV criticism.

That's perhaps not universally true of every single prime time serial of this period, but it's certainly true for a lot of them. As the decades unfold, and the relationship between television and televisions becomes more and more abstract, so too do the layers of self-reflexivity in these series. Shows that might have been about TV in the early aughts become, more broadly, about screen time in the teens. What these characters watch on their screens is an active part of how they understand themselves and how we understand who they are.

The Sopranos is about an actual gangster trying and failing to live up to the decades of mythology about gangsters he learned watching gangster movies and TV shows as a kid. (It's also about the corro-

Figure 2.3. The Kardashian children watching their dad on TV, *The People v. O.J. Simpson: American Crime Story*, "The Run of His Life," 2016

sive effect that watching too much History Channel can have on a sociopath.)

The Wire is about a group of people who don't have cable in their basement apartment, but, when they finally get it, they all become obsessed with watching whatever's on.

Homeland is about a woman who gets so invested binge-watching her favorite antihero series that she starts to sympathize and identify with its bad guy protagonist.

Halt and Catch Fire is about how TV antiheroes invented the internet.

The Americans is about two Russian spies who trick a CIA agent into believing that they're a normal American family by acting as if they are characters on a sitcom about a normal American family.

Scandal was precision-crafted by Shonda Rhimes and her writers to be live-tweeted. It's a show about Twitter designed to be watched on Twitter.

American Crime Story: The People v. O.J. Simpson is about what it was like for the Kardashians to watch the O.J. trial on TV when they were kids.

Stranger Things is about a bunch of kids who get attacked by VHS cassette tapes and get lost in a mirror universe that looks like TV static.

The Leftovers is a golden age television show that Damon Lindelof made about the very complicated feelings Damon Lindelof has about the public backlash to the final episode of his previous golden age television show, *Lost.*

Station Eleven is a dystopian miniseries about what it would be like to never be able to watch a movie or play a video game or search the internet or stream a dystopian miniseries ever again.

It is incredibly easy to find screenshots of people in each of these television shows watching TV themselves. It's almost as if they're doing it on purpose.

And then, there's *WandaVision.*

WandaVision is a mess of screens. The characters are on a TV show within their TV show. Actually, they're on several. And the characters on those TV shows within a TV show frequently watch TV themselves. The supporting characters on those TV shows are aware they are also on a TV show but powerless to resist the roles they play within it—they are, in effect, watching the show even as they are inside of it,

Figure 2.4. *WandaVision*, "Don't Touch That Dial," 2021

possessed by a power beyond their understanding. Various federal agencies, within the show, find the broadcast frequency for these TV shows, and *they* watch them—alone, and together—on a variety of vintage television sets. They pay obsessive, detailed attention to the show—a fan's attention—to figure out what's going on. Some of these agents literally write recaps of the episodes. In other words, in *WandaVision*, the lines between screen and viewer are violently policed and frequently transgressed. It's a show about TV history and media archaeology and the mind-controlling cultural power of a particular vision of heterosexual partnership and domestic harmony that spread through the viral medium of the television sitcom. It's also about a human woman falling in love with a piece of technology. About the possibility of conceiving children by touching a touch screen in the exact right way. And grief and ghosts and drone warfare and historical trauma.

So, in other words, it's kind of about streaming.

Before any of that, though, *WandaVision* is a show that only exists in relation to a movie franchise. It is one of the first narrative series developed as a spin-off of the Marvel Cinematic Universe for exclusive streaming on the Disney+ platform, and its relationship to that massive series of films is not a light one. Inasmuch as it's a TV show within a TV show, it's also a TV show within a movie, or at least *between* a few different movies, connecting them together.

In its first episode, we meet Wanda and Vision as the stars of a black-and-white, laugh-tracked, faux–*Dick Van Dyke Show* sixties sitcom. Wanda is Wanda Maximoff, an Eastern European superhero with the powers of telekinesis and energy manipulation, and Vision is *The* Vision, a super-intelligent and powerful android. They are in love.

But the episode does not acknowledge anything we know about these two from their multiple appearances in the MCU films that precede it, least of all the key fact that Vision is—oops!—dead. Importantly, *they* don't seem to know this fact either, and the episode passes almost entirely in the manner of the sitcom on which it's based, without any reference or explanation. In that first episode, called "Filmed

Before a Live Studio Audience," the show is clearly leading up to a revelation, some kind of gargantuan meta-twist, but even the slightest bit of speculation about what that twist might be requires that the viewer possess a tremendous amount of knowledge. Nothing about the show is even close to legible if we don't know who these characters are, and if we aren't aware of their bizarre and contradictory fates in the films of the Marvel Cinematic Universe. The bar for entry is at least three movies long, if not longer. Pulling off the greatest possible prime time serial trick, *WandaVision* requires us to have caught up with its story before the pilot even airs.

The tricks keep coming, as each of *WandaVision*'s first seven episodes is modeled structurally after a particular decade's most iconic sitcom form, and we slowly begin to realize that the TV world in which Wanda and Vision exist is a massive telekinetic biodome. A psychic soundstage. Wanda, it turns out, has mind-controlled an entire town in New Jersey in order to live in a suspended state of domestic bliss with her lost love. The sitcom structures, we learn, are Wanda's safety blanket, her way of dealing with the grief of Vision's death. She's hijacked a suburb in order to swaddle herself in the soothing hum of the TV set. It's a grim vision of screen time: the fantasy escape promised by television is actually a sinister mind-control prison with devilish origins dating all the way back to the witches of Salem. (Who, in this context, are actual witches.) Thousands of lives roasting in the flames of Wanda's electronic hearth. *Watch next episode?*

The rigid temporal forms of television bend easily to a screen time era viewer who understands their power over her. And that's something like the point of the show, if it's got one. Attentive viewers—hi!—will note that the first seven episodes, each tethered to a specific model series, do their best to adhere not just to the conventions of that specific show, but *to its run time*. The final two episodes, however, eschew the sitcom form, but they also balloon in length to the low forties. They become what the show was all along: a prime time serial.

To understand any of the subtext—which is really the text—of the sitcom episodes, the viewer had to be piecing together and following along a serial narrative taken up straight from the end of *Avengers: Endgame*. Viewers were necessarily constructing a parallel plot alongside the series itself, as a kind of reference timeline. The final two episodes merely acknowledge what the viewer had been doing all along. We were watching a series of classically structured sitcom episodes—which, for the most part, conclude their stories within the space of each episode—and nestling them within an ongoing narrative framework the show hinted toward but did not fully provide. (Placing each in its place, like, for instance, stones in a gauntlet.) We were serializing. And so was Wanda. At a certain point in the penultimate episode, Wanda's nemesis compliments her on the elaborateness of her creation. She says, "Thousands of people under your thumb all interacting with each other according to complex storylines, well, that's something special, baby." Sounds familiar.

Wanda's psychic reinvention of her universe is a good metaphor for the state of TV series on streaming platforms. Sitcoms on streamers often run around a half hour, and they often look like sitcoms from decades earlier. Same with prime time serials. But they don't have to. Freed from time slots and commercial breaks, they don't need to adhere to specific runtimes. They don't air at specific times. Any adherence they have to the old forms is merely a matter of tradition.

Or, rather, it's a matter of feeling: the intentions of the writers, the expectations of the viewers, their nostalgia for the shows they grew up with. Any fealty they show to the historical form of television is a fantasy projection. The connection this series—and the other Marvel series streaming on the same platform—has to TV's historical time frames are easily breakable. The show we're watching is as much a TV show as the one Wanda's creating in the diegetic world we see onscreen. These are the new constraints. There *are* no constraints. There is no television set. There is no cable. It's not TV, it's *WandaVision.*

Ultimately, *WandaVision*'s structure is a gimmick, an extreme version of the kind of "puzzle box" series that's become popular over the course of the new century. But it's a gimmick that works, a puzzle box to which we all have a key, because we know it already.

If the prime time serial is *the* form of the screen time era, then *WandaVision* is something like a diary of screen time consciousness. The anachronisms and media confusions between TV and streaming and movies and advertisements and life in your living room all make sense within the space of the show because the show itself is a projection of the screen time mind. (In her book *Millennials Killed the Video Star*, Amanda Ann Klein provides convincing evidence that reality TV series are another perfect form of the screen time era.) I can't relate too much with Wanda Maximoff's coming of age in war-torn Sokovia or with her grappling with the implications of being a powerful necromancer. I can relate a little bit with her obsession with a computer, and certainly with her nostalgia for VHS tapes of broadcast television series. But I know exactly what it's like to see the world through a mess of screens, to feel time open up or gain structure in relation to them. I know what it's like to live in screen time, too.

I'm just a couple years older than Wanda Maximoff, so, despite growing up in pretty different circumstances, the Scarlet Witch and I still have a lot in common. Like her, I can remember a time when screen time could only mean what transpired on my wood-paneled RCA Colortrak TV. I remember when computers entered the picture. I remember the first time I fell in love with one.

But Maeve, who was born in 2015, deep into the screen time era, understands it differently, and so does her media. *iCarly*, for instance—the oldest show Maeve watches, from 2010—is about two teens who inadvertently start a super-popular webseries. *That Girl Lay Lay* is about a teen whose avatar from an affirmation app comes to life and moves in with her. Some of these shows that she loves are like *Twin Peaks* and *Mad Men* in that they take the characters' constant mediation of their own lives by way of screens as a central subject. But, you know, for kids.

That Girl Lay Lay, especially, which—like *WandaVision*—focuses on the relationship between a girl and a hyper-intelligent AI, doesn't just breeze over the bizarre, but bizarrely understandable, human/nonhuman relationship at its core. Screen time is a kind of time, and it's a kind of feeling. But the reason it can be either of those things is that it's a relationship, too. Maybe it's an unbalanced or unbalancing one, maybe it's an exploitative one, but it is one.

WandaVision and *That Girl Lay Lay* occupy semicomic dystopias about this very thing. They work, adjacent to the sci-fi genre, on the questions *Mad Men* asked through its historical realism: what does it mean to have a relationship with a screen? What does it mean to communicate through these sorts of alienated media? What becomes of intimacy, love, grief, friendship when we view them, often exclusively, through screens. *Mad Men* makes this its subtext, *WandaVision* makes this its text, and *That Girl Lay Lay* takes it as its everyday. *iCarly* and *That Girl Lay Lay* are geared toward those kid subjectivities who've never had any need to grapple with this as an emerging historical phenomenon. This is what it's like, and this is how it feels, I guess, to be a "digital native."

3

WE ARE NOT ALONE (GOOD VIBES)

In 2015, a lot of things about my personal relationship to screen time changed. Namely: *Mad Men* went off the air, and, then, five months later, my daughter Maeve was born.

I know *Mad Men* is not as important to everyone as it is to me. It keeps coming up in this account of screen time not because I think it's the only show that matters, but because of how intensely it's mattered to my own experience of screens and times. One of the ways that's true is that, for its final three seasons, I recapped the show with my friends for our blog *Dear Television*. When we first started, in 2012, we wrote as if we were correspondents in an epistolary novel. No recap was ever done alone. In those early days, we'd each contribute to a rotating chain letter about every episode of whatever we were watching. Me, then Lili Loofbourow, then Jane Hu, then Evan Kindley, writing with each other and to each other. Each letter addressed, "Dear Television . . ."

The four of us were purely internet friends, at that time. Having connected on Twitter, arguing with people about the TV shows we were watching, we'd never actually met each other in person, in the meatspace. Our relationship was born and grew up in email, text, Twitter DMs, fathomless scrolls of GChat (RIP), and, of course, on the blog. These people I knew only online would become, and still are, some of my most intimate interlocutors and collaborators and, eventually, friends. *Creatures made of screens writing about screens.* For every week of TV, we'd produce anywhere from seven to twelve thousand words total between the four of us. After a while, because we left the endless scroll of the Wordpress for the slightly less indulgent pages of an actual publication, and because that much TV criticism was unhealthy to write over a period of days, we started working in rotating pairs. Eventually, after recapping a bunch of different series on the blog and for other platforms, we landed at the *Los Angeles Review of Books*, where we were when it fell to me and to Lili to recap the series finale of *Mad Men* in 2015.

As we know, *Mad Men* is a TV show about a bunch of different people looking at each other and then looking at their TVs, and one of those people was me: at 10 p.m. EST every Sunday, nestled into my couch at the convergence of all those looks. And looking at all of those screens.

When it was my off-week, of course, I'd watch just with Mel. Like a lot of people, we had a standing appointment to watch whatever was on on Sunday nights. Our favorite shows in that time slot were the speedy procedurals of *The Good Wife*, the maddening clusterfuck of *Game of Thrones*, the rollercoaster romances of *Insecure*, and recently, the nineties nostalgia puzzle box of *Yellowjackets*. It was rare to have a Sunday without a big screen time era event series to turn on.

These are what used to be called "water cooler shows," or shows that people would be excited to talk to their colleagues about at work the day after. The idea of the water cooler show or moment emerged back in the network era, when there was simply less TV, and so more viewing experiences could be more broadly shared.

But Twitter has helped to rescue the networked-era water cooler, bringing it into more of a work-from-home situation. (Shonda Rhimes, recognizing this, made social media engagement a huge point of emphasis in developing her series, and writers from the shows of the Shondaland empire still regularly muck about with fans online while their episodes air.) Twitter moved the water cooler discussion online, where you and everyone else who watched whatever you watched could talk about it, even if nobody at your actual physical office the next day would have any idea what you were talking about. There, you were online with your co-workers in the culture industry.

But, for me, before any shocked reactions or eviscerating recaps could occur between me and my friends on the internet, there was Mel and me, sitting on the couch, together. Unlike my compatriots in *Dear Television*, Mel *is* someone I met in the meatspace. But our relationship, like a lot of relationships in this era, has a long screen history, and was lived in part *through* screens. Not only, and not exclusively, but often. We started flirting in person, in the graduate computer lounge of the University of Pennsylvania's Fisher-Bennett Hall, speaking to each other for months from behind rows and rows of Apple desktops (screens). But it was only after the flirting moved to email (screen) that I, a dope, recognized it as such. The emails built, and then we started flirting in text messages (screen). I didn't really know how to proceed, so I texted (screen) our mutual friend Dave, who I knew was planning on watching the Philadelphia Phillies play in the MLB playoffs at a local bar with Mel, and I scammed an invite from him. Not being a baseball fan at all at that moment, but being very invested in appearing to know what I was talking about, I spent hours googling facts (screen) about that year's team. So our first couple dates transpired in Callahan's on South Street, watching on multiple flat-screen TVs (screen) as the Phillies bulldozed their way to the World Series in 2008. Me, Mel, Dave, Mel's brother Ian, and dozens of other people. All together, watching, water cooler style.

That was the beginning, but screens stayed in the picture. We held an annual Oscar party at our apartment (screen) that required guests to make thematically punned desserts. We spent hours upon hours reading and editing each others' book drafts in Microsoft Word (screen). We watched the final three *Harry Potter* movies with a bunch of little kids at the IMAX theater in the Franklin Institute (screen). We FaceTimed (screen) every day for the two years when I lived in Baton Rouge and Mel lived in St. Louis for work. When we first moved to the Midwest, we bought MLBTV (screen) so we could continue to watch Phillies games and feel closer to home, and closer to that slightly easier early time in our relationship. We eventually subscribed to NBA League Pass too (screen) so we could watch the Sixers, until we grew tired of the glitchy and expensive official subscription and began to stream games illegally through a creepy, if reliable, site called BuffStreams (screen). When I was away from home (and from cable) for the 2014 Emmys, Mel set up a laptop (screen) to Skype me our TV set (screen) so I could watch live. And Mel texted me a photo of the positive pregnancy test when she found out she was pregnant with Maeve (screen)—my cat photobombed the picture.

Every semester when I teach my intro to visual culture studies class, I ask students a question, adapted from two essays by Susan Sontag and Lauren Michele Jackson: "Is the internet—the world of images—real life?" Students have a few minutes to think, but every time, more answer in the affirmative than not. They quibble about the definition of the word "real," they point to the internet's capacity for misinformation, its status as a platform for lies, but they nearly always come back to the unavoidable fact that real things, meaningful things, things that have affected them in material ways, things that have verifiable existences, have transpired online. For these students, to say that the internet isn't "real life" is to say that a substantial amount of their own lives isn't real. Maybe the answer is a coping mechanism, an intellectualized cover for the ephemerality of life for digital natives. But I'm inclined to believe them, because it feels true—maybe bad, but true—to me, too.

And so it is with the screen time I've spent with Mel over the course of our relationship. It is not insubstantial. These moments are real life—or at least they are adjacent to it—and they occurred through, next to, sometimes because of screens. This is true for lots of people, even people who might feel bad about it. And ritually watching our Sunday shows is one of these things.

I shouldn't need to explain this part. Whether it's watching a movie in a theater or on a hillside or watching a favorite show on the couch, to watch together is a particular form of intimacy. Child psychologists have long said that the best way for kids to experience screen time is to have a parent watch—co-view—with them. The easiest way for children to be educated in any way by "educational" programming is to view it in rolling conversation with an engaged adult. The byplay between the kid and the screen complemented by the byplay between the kid and the parent or caregiver. The twelve-to-two slot on Sunday afternoons—Phoebe's naptime—is a time we've always set aside for co-viewing with Maeve. We watch a movie or a show we're all invested in, and we take a rare opportunity to do what the child psychologists would want.

The value of co-viewing is portable to the time after the kids go to sleep. When it's just Mel and I, we can talk to each other, we can talk back to the screen, we can, for an hour or so, live our relationship—this complex, detail-filled, sometimes tense, often romantic, long-running narrative—in and through our Sunday shows. It isn't everything, but it's not nothing.

But that was all on my off-weeks. Every other week, I was watching with the *Dear TV* gang, and I was looking at a lot more screens than usual. There was my TV set as the episode aired; my TV over and over as I rewatched through the night; all the tabs and windows on my laptop; the Word document where I transcribed quotes and notes and slowly built paragraphs upon paragraphs of criticism; my email inbox, where my friends and I talked about what we were thinking and writing about; texts on my phone; Twitter, where I read other fans and critics responding in real time; the screengrabs I snapped and pasted into the CMS of our website; Wikipedia, where I looked up historical details; other recaps that dropped before mine by critics I admire; Facebook, distracted.

Ordinarily, recapping was not necessarily this kind of endurance sport, but AMC, as a matter of principle, never sent out advance screeners for *Mad Men*. So when it was time to recap the show, that meant that a particular ritual was about to commence: plopping down on the couch at ten, doing one clean watch with Mel like a regular human being, saying goodnight to her, rewinding the DVR, doing a rewatch while taking notes, rewinding, keeping the show on in the background while I drafted, rewinding, emailing, drafting, rewinding, and pausing to get lines right, pulling screenshots, and sending three thousand words to my editor at 3 a.m. EST. If all that worked out, we'd have the piece up Monday morning, in time to catch the wave of viewers waking up to seek out their favorite recaps. And I would slink to bed and wake up the next morning, capable only of periodically checking in on how the piece was doing on the site. One night and one day devoted to the sacred duty of proving I'd been paying attention to a prime time serial worth paying attention to.

Mad Men went off the air in May 2015, and, five months later, Maeve was born. That was both the beginning of my parenting life and, not coincidentally, the functional end of my recapping life. This ritual was how my screen time worked before my kids came along, but it's hard to keep that rhythm alongside the rhythms of parenting, for me, ever the moonlighter. Writing about TV, even watching it, still carries something of that time-deranging quality. I write in contained sprints, and I stay up late to do it. I take notes, even if not on paper. I anticipate the rewatch that will inevitably follow. I still recognize that time signature that I only know from those days staying up late, locking eyes, with Don Draper. Keeping my appointment.

* * *

Beyond the TV set or even the commercial break, you'd think that the real endangered species in the screen time era would be the appointment. It's the idea of sitting down at a particular time every week to enjoy your favorite shows—your stories!—that one would imagine to be most imperiled by Netflix's patented full-season dumps, the wide embrace of on-demand media content, and just the sheer *availability* of media in these years. Streaming search engines allow us to be capricious, and algorithms allow us to impulse-stream. The binge-watch—plowing through episode after episode, hour after hour, getting up only to pee and fetch snacks—might seem akin to the appointment, but it's something else entirely. The binge is unregulated and excessive, while the appointment's pleasure is precisely in its regularity. Both tether us to our couches, but the binge is a flouting of TV's historical and temporal logic. It's about control, too, just differently. Even still, binge-watching has not killed appointment TV. At least not yet.

Since the invention of the VCR recording function, it's been possible for viewers at home to unmoor what airs on television from its air time. It should not surprise you to learn that, along with Tom Engelhardt's manifesto and the prime time serial, TiVo was also a product of the screen time nineties. The ability to pause live TV, to record broadcast episodes, and, eventually, to order up previously broadcast series via cable menus changed the way a lot of viewers conceptualized the urgency of the time slot. And it also opened the way for streaming platforms to structure themselves as on-demand services as well. You could go back and catch up on series that had never aired at all. This type of screen menu—where all options exist simultaneously—implies an eternal present. But, for those of us who grew up with firm TV schedules, everything we find on demand feels somehow out of time.

Appointment viewing for kids is both incredibly fungible and necessarily routine. Within the home, "screen time" can be a substance to control, but it's also, always, a tool. Maeve and Phoebe need screen time to survive our cross-country road trips; I need them to have screen time so I can cook dinner; Phoebe has screen time to relax before nap; they have it withheld if they demand it too sharply or seem at risk of wilting in front of the TV on otherwise lovely days; it's a good way to get them to slow down for a minute and cuddle with us. It's required, regularly, at certain points of the day, but it's also free-floating. Growing up, my screen time was often tied to what I wanted to watch: *when's the DuckTale start?* For my kids, though, screen time is tied to times of day, and nearly always revolving around on-demand content. My appointments were for shows; their appointments are for moments. The networks used to be in control. Now I am. Sometimes.

In 1981, a theologian and art historian named Gregor Goethals published a book entitled *The TV Ritual: Worship at the Video Altar*. The argument of the book was a not unfamiliar version of a common secularization narrative, chiefly that television was in the process of replacing—and, of course, degrading—religion. TV's ability to produce virtual community, satisfying a human need for heroes and pageants that used to be filled by religious ritual, threatened the underpinnings of religious belief in America. Goethals's argument proceeds pretty literally, analyzing TV broadcasts of actual rituals: the Super Bowl, the JFK funeral, royal weddings. Forty years later, television—now a well-established religious system, with video altars in every home—is undergoing its own secularization crisis.

But its rituals remain. Whenever it seems like it's time to start tolling the bell for appointment television, an old series returns or a new series takes hold of its must-see excitement. *Game of Thrones, How to Get Away with Murder, Succession, Power, Better Call Saul,* even network sitcoms like *Abbott Elementary*. And they remain because, like the religious rituals Goethals imagined them to replace, appointments give us something. They give us screen time that connects us with other people, they give us screen time that we can look forward to—rather than the compulsive screen time we get from fidgeting scrolls and habitual pocket pulls—they give us screen time that transcends, even if for a moment, the screen we're watching or the program that's airing, they give us screen time that feels good.

* * *

When we drive to Pennsylvania in the summers, with the girls locked into their screens for the miles and miles of cornfields and blasted out hillsides, we drive there to visit the relatives we left behind. In the parlance of our times, we take these trips for face-to-face, or f2f, contact. For Maeve to rustle her Gram's many German shepherds, for Phoebe to climb on her Grandpa Foo's back, for the both of them to fall into a real pile with their Uncle Ian and Aunt Lolo. But, for the vast majority of the year, Maeve and Phoebe and their Philly family talk on FaceTime.

It's very difficult to understate the degree to which I specifically did not believe that video phone technology would ever be a thing. Like a lot of aspirationally pretentious suburban teenagers, I went through a period of twee Luddism in the late 1990s. Inspired by the Beastie Boys, I bought dozens of vinyl LPs for ninety-nine cents a piece, I made a cut-and-paste zine about indie music called "The Electric Soul Potato[e]" with my friends, I asked for and received a manual typewriter for Christmas. These were the broad trends of the thrifted-cardigan-over-gas-station-attendant-shirt-wearing white boys in my demographic, but my analog aesthetic was, for a time, animated by a genuine pessimism about technology in general. Partially as a stylistic choice, and partially as a real belief, I remember very casually talking about the silliness of striving toward things like voice activation, digital navigation, and, importantly, video phones. In the nineties, my vision of the future was one in which millions of dollars would be spent trying to perfect marginally useful *Jetsons*-inspired technology that would never ever really work.

It only now occurs to me that this popular culture of tech backlash, of which I was a teen devotee, was itself a phenomenon of the screen time era. The teenagers who brought their antique Olivetti typewriters to coffee shops to write Vonnegut-esque short stories are the same teenagers whose youths were the first to be governed by this particular parenting movement. These are the kids who were told screens were bad for them, who had TV banned, or who overindulged in response. Though I doubt anybody in this group would have listed *obedience to parents* as a particularly high priority, it strikes me that at least a part of this allergic reaction to slick digital technology—technology that Apple was making slicker and slicker by the day in ways that would eventually tempt us away from our tech-free purity—was about having grown up within a cultural moment defined by the villainization of screens. *Maturity means the ability to discern.*

But I was wrong, it turns out. FaceTime works. Or, rather, the technology of FaceTime works. The user experience can be a little buggy.

There have been several stages to the girls' use of FaceTime. The first stage was the easiest. The child—Maeve in this case—is a small, swaddled dumpling. Mel could call her mom or her sister and, magically, have an ordinary conversation, with a live feed of Maeve onscreen instead of her own face. What if I told you that you could talk to your own daughter but see only an uninterrupted video of your infant *grand*daughter? The future is now! This is the excellent deal that Gram cut in those early days. But then Maeve got squirmy. That period has lasted from a couple months old, roughly, to the present. In the first stages, that squirminess caused a troubling wrinkle in our FaceTime dynamic: we couldn't keep Maeve onscreen. We could talk, but anyone on the other end was watching a feed that looked like somebody had thrown a GoPro camera down the stairs at an IKEA. These FaceTime conversations were less rewarding than the original ones, and they'd invariably end with a loud offscreen noise and a hurried "look, I have to go . . . love you" before the feed went cold. FaceTime was working, but we simply could not offer the stability—in terms of focus or even just physical motion—required to get it to work for *us*.

Then Maeve ascended to late toddlerhood. She was still squirmy, but with better motor skills and a pliant, inquisitive mind. At that point, the paradigm shift occurred: we just fucking handed her the phone. Her framing instinct was not fully developed just yet, so often these images consisted of the top of her forehead in the bottom of the screen, a roving shot of our ceiling fan, or perhaps just a close-up of her nostril. But, without indulging in too much age-ism here, her grandparents were not all that much better. This was especially true of her GG Pap, *my* grandfather, who was still around and always eager to pick up his iPhone when Maeve called. (Even now, months after he passed, his contact is listed in my phone as "iGrandpa.") One of the most enduring images I can conjure of him is of a four-year-old Maeve gabbing jubilantly about nursery school while holding a phone that showed a screen image of my Grandpa's right eye with an inset image of Maeve's right eye. Looking out, looking in.

With an older, wiser, calmer Maeve, and a fidgety toddler in Phoebe, the FaceTime situation has become somewhat untenable again. Mel holds the phone as the two children rocket around each other. It's mostly Mel in the frame, looking apologetically at her conversation partner, hoping one of the children inadvertently zips into view or spontaneously discovers the concept of guilt. Otherwise, the image our relatives see in Philadelphia is mostly akin to those deep space images that tell us about what we can't see by showing us how what we *can* see behaves. This is what it's like, our FaceTime screen tells them. This is how it feels.

In other words, FaceTime has never really not been stressful. But that doesn't mean it hasn't been a miracle of sorts. Its mere existence closes distance, the promise of it does a lot of emotional work, even when the actual experience is wanting. The payoff is answering the phone and *seeing* the person you love. The payoff is making the call at all, anticipating that face. Its utility lives in that split second; everything else is gravy. It isn't a replacement for contact. The screen is not usurping physical closeness. It will never do that. It *could* never do that. But it can offer something else, something in the neighborhood. Perhaps because we've learned to build these relationships with screens—with characters we love or hate, with events we've anticipated—we know how to have intimacy through them. It isn't the same as person-to-person, but it isn't nothing, it isn't cheap, it isn't degraded. It is simply something else on its own.

The first COVID lockdowns began during Maeve's spring break from nursery school. We never sent her back. They put together some cursory online meetups over the remaining several weeks of school, but it's not like they had any curriculum they needed to finish. The kids all squirmed in their seats while the teachers sang songs for half an hour, and then everyone logged off.

The following fall, though, Maeve started pre-K at a real elementary school, and that school, we are grateful, was fully online. We pulled up to a drive-in circuit in the school's parking lot that August, picked up a box of worksheets and supplies, and checked out Maeve's own personal iPad, provided by St. Louis Public Schools with a chunky little purple case. Her class met every day on a byzantine schedule laid out—mostly accurately—in a video chat app called Microsoft Teams. They'd meet first thing in the morning for songs and alphabet and show-and-tells and several rounds of explanation about how to mute themselves, then log off for a worksheet, then back on for the subject of the day, then off for lunch and rest, then back for science or reading, then the day was done. Because Mel and I were snowed under managing our own online courses, my mom took over as something like Maeve's schoolday concierge. She'd sit next to her when she was logged in—just out of frame—helping to keep her focused. She'd usher her in and out of virtual rooms. She'd help her with her worksheets when she needed it. She was Maeve's preschool teacher, and she was wonderful at it. Microsoft Teams was not a replacement for school, but it gave Maeve, and my mom, enough to work with.

In the spring, when the school doors finally opened, Maeve returned. There'd been a lot of doomsaying about learning loss due to online education. And there's no doubt that there was indeed something lost between these kids and their teachers that couldn't be communicated on an iPad.

But it's also true that nothing truly bad that happened to these kids happened because of a screen. A new virus spiraled across the planet. Family members, friends, teachers died. Businesses shut down. The federal government chose to prioritize bars and restaurants over schools that first summer, and so schools shut down, teachers quit. Parents and educators were stretched to breaking points because a system nominally designed to support them simply chose not to.

We were lucky, we know. Kids who were handed screens without the family and school infrastructure we had were handed nothing, essentially. The screens did not save everybody, but that was never a thing that was in their power. As with masks, it's easy for people who feel the world slipping out of their control—as we all did—to imagine that it was the (insufficient, frustrating, buggy) *solution* to the problem that deserved the blame. Maeve's screen, just like her mask, didn't do anything but help keep her and her friends safe for months and months. The world collapsed on these kids, but screens, it turns out, did not.

And that was okay. Screen time is not nearly as alienating a communication medium for Maeve and her friends as it is for all the middle-aged pundits who decried virtual school as an abomination. There are things these kids want that are physical and material and "in person," but they are growing up in a universe where screens are capable of doing things like this, and where intimacies exchanged in passing on them are not second-order or fundamentally degraded.

The moral panic about virtual learning is about what all the other moral panics are about: growing up. This is a growing up that's not just worrisome because of the loss of time and childlike innocence and closeness it implies. What does it mean for our children to grow up *different* from us? Different technologies, different classrooms, different traumas—the things that seemed real to me when I was growing up might not seem real to them. The things that seem real to them seem unreal, ghostly, to me. To raise kids in this particular screen time is to feel the constant, terrifying tug of one's own obsolescence.

Maeve finished kindergarten this past spring. Her year there was back and forth—some virtual switches, some mask-on/mask-off guidance, only one outbreak, from which we were mercifully spared. Somewhere in there, she's learned to read at a high level, and she does so avidly. Mel—who was an early and avid reader like Maeve is—had often talked about the small, sweet dream that she'd one day get to sit in a room with her young daughter as they each silently read their own books in each other's presence. That dream is real now, and it's in part because of the screens—and the humans like my mom and like Maeve's pre-K teacher who kept them in the right place—that precisely prevented her learning from being lost.

Early in the pandemic, we got a retro, corded landline telephone in our living room. It's Maeve who uses it most. We gave her a list of phone numbers—

all her grandparents, her aunts and uncles, a few other folks—and the rule is that she can dial any of those numbers whenever she wants. She sits in our little green armchair, calls up her Gram and regales her with tales of the day, calls up her Uncle Ian and asks him about the bear stuffies he keeps in his work-from-home office, calls up her Aunt Lolo and reads her complete recipes from a cookbook for some reason.

She's not addicted to screens; they didn't take anything from her. The face time provided by FaceTime isn't enough, but neither is the voice time provided by the phone. Screens weren't ever going to fix that. We shouldn't have asked them to. The problems we have with screens are often problems we have with the world screens exist to mediate and capture, imperfectly, for us. They can't close the distance, they can't bring us together in the same room, they can't fix a pandemic or teach a child how to read. Screen time can't do it. There isn't enough time; no time is enough.

* * *

One of the things my students most frequently cite as evidence for the reality of the internet is social media. They mean this in positive terms—as in, the internet is a place where friendships form, romances kindle, communities thrive and organize, and families reconnect—as well as negative terms—as in, the internet is a medium for hate speech, bullying, the policing of bodily norms, sexual norms, gender norms, all the norms.

Both the positive and the negative go to the same point, though, which is that social media allows for the internet to become an extension of real life, a place where real, on-the-ground dynamics can extend across geographical and even cultural boundaries, absent only our bodily presence. I'm not sure I know a better example, or more depressing cautionary tale of this, than Vine. Its life and death are, in some ways, the story of the screen time era.

On October 28, 2016, Twitter cut Vine. The death of the popular social media app, which had allowed users to shoot, edit, and share six-second looping videos with shocking ease for three years, was met with dozens of online obituaries, heavy with embedded video content remembering joyful moments past. Jeremy Gordon memorialized it as "the only good app"; Rembert Browne called its shuttering a "hate crime against black teens from Atlanta." These essays and remembrances, much like the many appreciations of the app produced while it was still active, made a collective case for Vine as a signal work of art in the age of screen time. What it did with attention, what it did with community, what it did with open access—Vine imagined a way out, through the oversaturated screen time consciousness itself.

Formally, Vine was the perfect, portable genre of the screen time era, capitalizing on its desires and anxieties just as well as the prime time serial did, but innovating around its edges too, moving the form toward viewers in ways that the antihero shows of premium cable only ever did metaphorically.

Vine was an app for screens watching us watching. The very act of viewing was embedded in the form. The record function was physical and responsive. The very first films ever made at the end of the nineteenth century were a lot like Vine videos: short, sometimes shocking, sometimes mundane, always focused on their imagined audience. And they tended to fall into two genres: the trick and the actuality, the spectacular and the everyday. One of my favorite Vine videos is of a man falling out of bed. The second he hits the ground, his body transforms into hundreds of marbles that scatter across the floor. Watching it on repeat, the shock of the transformation never, ever gets old. Simple, but seamless; the magic of the ordinary and the magic of the camera. And it's done with one of the earliest innovations in cinematic editing, the same trick Georges Méliès used to make ladies disappear or Moon men combust in the 1890s: the substitution splice.

It felt good to edit video this way, transforming the laborious processes of early cinematic tricks into something intuitive for beginning users and tremendously freeing for artists. It was a form easy to learn but easy also to use as a jumping-off point for exploration. The scholar and critic Ashon Crawley described Vine as having an "aesthetic of possibility." That possibility existed, at root, because all of Vine's creators were viewers first, and all of its viewers accessories to these acts of creation and possibility.

But the act of watching was not just a part of these videos because they existed on an app where all viewers were also creators. The act of watching floated at the surface of the Vine experience because of the *way* the videos were presented. They focused, in ways that mirrored some of cinema's earliest moments, on what *drew* viewers in the first place. What gestures, what scenes, what elemental essences would grip a viewer in a short space and time? Sometimes that was about the way a particular bizarre image struck the eye, but sometimes it was about deep referentiality, isolating and remixing a sound or an image so viewers gravitated toward it as familiar, even uncanny.

Great Vine videos could produce an image that seemed briefly magical, a nearly impossible feat in a culture defined by the ubiquity of easy-to-use digital technology. But they also focused—again, in ways that echoed early cinema—on what could *keep* viewers: loops. Vines looped. Over and over, without end. Some Vines are great on a single view, but the best of them, the ones that truly understood the form, grew into, through, and out of the loop.

Maybe the most archetypal Vine is a video that's about a Vine video being made. Uploaded by user Diamonique Shuler in January 2014, "I Ain't Gonna Do It," like any Vine video, is almost impossible to describe in writing. That isn't because its image beggars language; it's because what it is, what nearly all

Figure 3.1. "I Ain't Gonna Do It," Diamonique Shuler, 2014

good Vine videos *are*, exists in the act of watching. The video, simply, shows two figures—a young girl in a leotard standing in front of a mirror and a young woman, seated. The woman chants, "Do it for the Vine." The younger says, "I ain't gonna do it." The older repeats, in rhythm, "Do it for the Vine." The younger replies the same way again. Then, for a third time, the woman says, "Do it for the Vine," while drumming on her chair, and the young girl picks up that rhythm and dances her answer, "I ain't gonna do it." And then the video's over. Or, rather, then it starts again.

Crawley describes the Vine's finale this way:

> Disrupted after the third request, the girl in Diamonique Shuler's Vine loop begins to dance more emphatically, producing the otherwise. With what seems to be liberating happiness, unfettered and undeterred joy, her subsequent dip-head-sway-hip forces a reflection backward on the whole of performance, how its entirety is within a tradition of black performance, of black joy. But how?

The looping incantation of the Vine produces what Crawley calls "the otherwise," some extra, transcendent content, some space deeply within but also beyond the bounds of the six-second video. And it's a space produced through attention. The anti-screen prophets warned of viewers glued to screens, unable to pull themselves away. But what, Crawley seems to ask, if that attentiveness could be a vehicle? What if falling into the screen was a movement toward something beautiful and rare?

Crawley's exegesis of this Vine brings two of the app's screen time era signatures into focus. First, it transmogrifies attention, turning attention itself into a creative act. And, second, it joyfully resists the isolation that screens seemed to portend. Vine was social, but its sociality was not just about circulation and viral spread. As many of the obituaries wrote, Vine had become, almost immediately after its invention, a hub for creators of color. Indeed, Jazmine

Hughes wrote that it had "become a home for young black people, who dominated the service and established its visual language early—quick cuts, referential jokes, deep allusions."

What Vine owed to early cinema and even to the screen time era complexity of contemporary TV, it owed to Black creators. As Doreen St. Felix wrote in her oral history and tribute to six such creators—Jay Versace, Kelz Wright, Khadi Don, Landon Moss, Victor Pope Jr., and Yung Poppy—"young black Viners became internet-famous on the platform and beyond for their uncanny, refreshing takes on the canonical styles of black comedy." White viewers might not have perceived these foundational influences—especially as catch phrases and dance crazes from the app were virally appropriated beyond it—but histories of Black comedy gave shape to whatever it was Vine became.

But Vine coexisted, as well, with another genre of viral video. In 2014, Michael Brown was killed by a police officer in Ferguson, Missouri. His death sparked the nationwide Black Lives Matter movement, but that movement was catalyzed by the increasing availability and circulation of videos of police violence. These videos, formally, were not unlike Vine videos, and not unlike their early cinematic forebears, too. Short, shared, shocking—with the shock of violence, rather than joy or surprise—and looped. And they represented disparate, but deeply related, visions of Black visibility in America. It was not a coincidence that Vine's medium for Black joy and expression existed in counterpoint to this other medium of Black pain and death and mourning. For Crawley, this similarity allowed Vine, in its fleeting moment, to use its possibility to fight back, to "repurpose scenes of violence."

While the movement for Black lives has grown and spread, Vine didn't last. Twitter—who'd purchased Vine to make it a visual analogue to their main app—never figured out how to make it make enough money or garner enough attention. (The popularity of Snapchat helped lead to corporate dissatisfaction with Vine's performance.)

Regardless of its profitability, though, Vine had begun failing its creators of color almost as soon as they revolutionized it. As Hughes notes, Vine worked like a lot of viral popular culture, all the way back to the nineteenth century: "rooted in the black community, then sucked into the mainstream where businesses benefit from them financially." Black Viners created value, and then other people sold it. Vine rarely featured its Black creators on the app, its structure allowed reposts that essentially removed authorship from creators the more popular their videos got, and the app never evolved to help creators monetize or own their own content. In fact, the company famously refused the demands of a group of creators who organized for both remuneration and protection from harassment. "We live and die by the internet," Lauren Michele Jackson writes, of Black viral stardom. "The internet asks for more." Many of Vine's strongest creators fled to Instagram or other platforms well before Twitter called it quits on the app.

Since Vine, other apps have emerged to take its place. Notably, TikTok has filled in part of the empty space. If anything, that app is even more animated by the self-reflexivity of the screen time era. If Vine was built on remixing and reposting, TikTok is built on literal scenes of spectatorship, creators split-screening themselves watching and improvising with other TikTok videos. But TikTok has learned the corporate lesson of Vine; better at surveillance, better at algorithmic manipulation. Vine's contingency, its impossibility, made its end so spectacularly sad and foreboding—a homespun supernova reminding everybody else on the internet about the fate that befalls every star eventually, especially those that burn so brightly. The companies that provide screen time for viewers are leveraged against precisely the type of screen time that viewers are hungry for. For the act of creation to begin on the other side of the screen, for viewers to look back in this way, is unsustainable. These platforms evolve to better support the real lives of their owners, not the real lives of their users. What they don't need becomes unnecessary, what survives and thrives without them, in spite of them, no longer has a home on their internet. Online communities, online art practices, online relationships, my students insist, are real communities and practices and relationships. But then, sometimes, with a shocking splice, they disappear.

4

WE ARE NOT ALONE (BAD VIBES)

In June 2021, I became obsessed with a video of an unidentified flying object, or what's now called an "unidentified aerial phenomenon." The video, which circulated first from a Twitter account called @Today_China and was later picked up and elaborately dissected on the r/UFO subreddit, was taken from a rooftop in downtown Shanghai. The phone camera points up at the night sky, brownish clouds

Figure 4.1. Unidentified, June 2021.

quickly passing overhead, and, as a variety of voices pitched at various states of alarm or awe rise around it, we begin to see the shape of an enormous black isosceles triangle, hovering over the city, negative space amid the clouds. It's an extraordinarily unsettling image—the found footage laundered through the horror genre, emerging again on my feed—as the apparent object floats seemingly close to this bustling urban scene.

Maybe as unsettling as the size of the object itself is its blankness. Unlike other amateur UFO videos that have circulated in the past, there are no lights, no erratic movements or tremendous speeds, no dynamism at all. It doesn't dart at unfathomable speeds like the lights in the videos leaked from US Navy pilots earlier that spring. It doesn't dissolve or dance or change colors like the infamous Marfa lights in Texas. Nor does it produce an iconic silhouette like the flying saucers of old. The object is calm, featureless, which makes it somehow more threatening. You watch the video over and over in its loop, and the object gives you nothing. It is insensible to you, unbothered by you, *over* you, in ways literal and figurative.

And that's, somehow, what makes the anxiety—my anxiety—about images like these so sharp and nagging. It isn't purely a fear of alien invasion or a vast government conspiracy. It's about the way these objects appear to not care about us, or perhaps only care about us in the most condescending of ways. They don't want to be seen, but they're also *able* to not be seen. They can evade our eyes, our consciousness, our concerns. Screen time has always been, for me, about the relationship between the viewer and the image, the potentially nefarious intent of the television set transformed into something else in its communion with me. But this image doesn't care about that relationship. This object is alone, unidentified. And so, then, am I.

This is not my ordinary screen time, it's not my regularly scheduled programming, but it's also not totally extraordinary. The classic argument against media saturation is about kids' brains turning to goo as they stare slack-jawed at advertisements disguised as entertainment. But a particular corrosive vision of screen time—the one I most fear myself—is this addictive, obsessive kind that pops up in moments of acute stress or anxiety. The kind that keeps me up at night, the kind that distracts me from the people I love, the kind that casts shadows of my worst fears in the shape of alien craft. What if all of that attention and all of that analytical spirit and all of that intellect were sucked down into the void of this dark screen time? What if I'd turned the internet into the absolute worst version of myself, and I could peruse all the nooks and crannies of it for as long as I can keep my eyes open?

We just talked about some of the screen time forms that feel good. Appointment TV, FaceTime, Vine—these are types of media, types of viewing, that work with and against these cultural anxieties to seem uniquely pleasurable, even mildly virtuous.

But those cultural anxieties have also helped produce screen time forms that feel uniquely bad.

Mel and I are mostly unbothered about Maeve and Phoebe's current screen consumption. They watch, mostly, in moderation, and we at least nominally approve of everything they watch. But we can only really feel this way because we're still largely in control of their screens and what's on them. At a certain point, that will change. Maeve and Phoebe have a couple of play cell phones, both of which were produced by child education companies and both of which come with lessons in the alphabet and phone etiquette. They're "good" toys. But they both also look and function like regular cell phones. It's eerie to watch one of them occasionally slumped onto the couch looking at that fake screen—they don't even have anything on them—and to know that one day they may slump that way for real. One day, the phones they hold might teach them more nefarious lessons than the ones they are currently learning about the alphabet. Their screens will try to teach them how to feel about their bodies, what kinds of violence are and aren't acceptable, what counts as cool, what counts as ugly, what counts as a person. One day, in other words, we won't control their screens for them; some days, I wonder if I'm in control of my own.

I was primed for that UFO video, in both the long and the short term. Since I was very young, when I have been worried about something, I have tended to choose an external, spectacular fear to project all my local worries upon. I have very little natural ability to modulate my concerns about something—either I am not worried at all about it, or my worry grows immediately to a scale of fantasy that it effectively leaves the original worry behind, dwarfing it and rendering it manageable in comparison. In one way, this anxiety is almost strategic. It's hard to be worried about, say, the SAT, when you're also worried about a meteor striking the earth.

In other words, I'm a bit of a catastrophist. Don DeLillo, in *Underworld*, his novel about the Cold War hangover of the 1990s, wrote that "it's the special skill of the adolescent to imagine the end of the world as an adjunct to his own discontent." When DeLillo wrote that in 1997, the year I turned fourteen, a few years into the screen time era, he might as well have been writing about me, specifically. As an ordinary (white) teenager of the end of the millennium, I routed all of my mundane social, educational, familial, and cultural fears through the literal end of the world. I'm aware that this reads as a particularly privileged, bourgeois American sickness, a contemporary neurasthenia. American Nervousness, 2022. All the same, I worried about apocalypses religious and secular—ecumenical, I was, in my fears. I was drawn to their mythologies and their internal logics and weirded out by the presumption that powerful people knew about these things and chose to keep them secret—that presidents and popes were aware of the facts about the fundamental strangeness of our world, and the workaday ordinariness they produced and policed was merely a convincing cover story.

The catchphrase about extraterrestrials is that "we are not alone," but, because everything I ever read about their existence implied a network of powerful people(?) withholding the truth about the universe from me, the idea of alien life made me feel lonelier than ever.

In retrospect, it's fairly transparent to me that my fascination with UFO videos that summer of 2021 was less about what floated above us and more about what floated among us. This was the second summer of the pandemic, and the first during which we felt that it was possible to take our road trip home to Pennsylvania. Mel and I had been vaccinated that April, but the girls had not been yet. We were planning on being ultra-cautious, only seeing close, vaccinated family members, avoiding eating in restaurant dining rooms or doing any of the many fun indoor kid activities at any of our stops—a cabin in Kentucky, a shore house in New Jersey, Airbnbs in Pittsburgh and Philadelphia, a deeply off-the-grid farm house in rural Virginia—and the numbers looked to be in our favor anyway. But after over a year of virtual school and work, grocery delivery, hygiene theater, and isolation, leaving our house in St. Louis was a lot. I was doing fine, though. Or, rather, my maniac focus on UFO videos was helpfully absorbing all my anxiety about the trip.

And I had plenty of material. For years, scholars and advocates of these sorts of things had been working to normalize the idea that Earth was regularly visited by unknown aerial craft, that these encounters are and have been regularly reported, not just by wackadoodles, but by professional military operatives, and that investigations into these events ought to be known by the public and discussed openly. These activists—who came up with the acronym UAP in part to rebrand away from the aforementioned wackadoodles—have worked tirelessly to convince the public and government officials that just because this stuff is invisible doesn't mean we can't talk about it.

In June 2021, a few days after I saw that triangle video, the activists got a big win. The Office of the Director of National Intelligence released a report summarizing its internal investigation into eyewitness claims of UAPs, mostly from the Navy, from 2004 to 2021. The report acknowledges 144 accounts, 143 of which lack firm explanation, many of which have visual confirmation in the form of videos or photographs. Still, the report was disappointing to many in the UAP activism community. They were thrilled with its existence, and Congress's willingness to take it seriously, but there wasn't a lot of meat on the bone. Many of the videos described had been available for years, and the structure of the report heavily implied that, while all of these objects were "unidentified," they were unidentified because of the limits of our vision and technology, not because their identity was a world-shaking revelation. Yeah, these things are weird, the report implies, but not, like, *weird* weird.

So we return to the videos. I've taught in and out of film studies programs for the better part of a decade, and I've never encountered more committed, detail-oriented, imaginative analysts of film and video than I have on the UFO subreddit. The readers on those pages are alert and deeply knowledgeable about digital effects, lens flares, video distortion, image resolution, camera movement, and the limits of photographic technology.

They are obsessed with a conspiracy, but their writing about these images is not all that conspiratorial. In fact, perhaps because they've come up in a generation that deeply wants the possibility of extraterrestrial life to be taken seriously, they are incredibly rigorous. This board is not a place full of credulous rubes—or, at least, not as it relates to film analysis. Their standard of evidence is tremendously high. This is an old story. Defenders of spiritualism and mediumship in the nineteenth century were similarly scientific in their defenses, but they were also eager to debunk frauds and false mediums in their midst as a gesture of their own seriousness.

And the UFO community, as you might imagine, is replete with high-profile debunkers. Perhaps the most prominent of these debunkers, a British science writer named Mick West, nearly always begins by dissecting new videos *as* videos. His elaborate, fine-grained debunkings of most of these videos reveal, more often than not, the limits of video cameras to see what's right in front of them. Fighter pilots and Navy seamen are trained military professionals, but they are not professional photographers or camera operators or even film critics. What they see, often, is a phenomenon *produced* by the camera, rather than recorded by it. Working for the Navy is pretty stressful, I'd imagine. I can certainly relate to the experience of looking out at the vastness of the sea and seeing my own fears reflected back at me.

My occasional obsession with these videos is dilettantism compared to the investment of these activists and debunkers. Part of what's appealing to me about the genre of the UFO video is that it is made to be analyzed, it rewards the attention of these obsessives, it exists, like the prime time serial without the prestige, *because* of their attention.

And so, when I saw that triangle, I turned to them to tell me it was fake. And it was. It turns out that, in a city with as many skyscrapers and spotlights as Shanghai, it is possible for buildings themselves to create shadows in the sky. What those onlookers had seen was the shadow of the very building they were standing on. The movement of the clouds, gently grazing over the shadow of the skyscraper created an illusion of a hovering object—its blankness made sense because it truly was nothing, a shadow puppet of an alien craft that we'd all convinced ourselves was real.

* * *

When I was first able to read independently as a kid, I gravitated to the encyclopedic. The books I loved most were big compendiums of information. We had a leather-bound *World Book Encyclopedia* set that I used constantly, I loved atlases, I loved baseball card guides. When I got into music as a tween, without the aid of Spotify or even too many cool friends (or older siblings of my uncool friends), I turned to big reference books again. *The Rolling Stone Album Guide*, *Rock Movers and Shakers*, and *The Spin Alternative Record Guide* were in heavy rotation. When I started to style myself as a writer, it was *The Salon.com Guide to Contemporary Authors*. I loved reading short entries about things I hadn't yet experienced.

I read deeply, I cross-referenced, I memorized dates and chart positions, I created a vivid sense of what I thought R.E.M.'s early single "Radio Free Europe" sounded like before I had a chance to hear it, I came to hard and fast conclusions about what the best Replacements records were or which Fleetwood Mac albums constituted their decline, I knew who Sonny Sharrock and Captain Beefheart were—and respected them!—without having heard a note of their recorded output. I loved the freedom and impulsiveness of this style of reading, the feeling that I could read at a more and more energetic pace and that, by the time I was done, I would have learned something specific, come to some sort of satisfying knowledge, even in aggregate. But I also loved knowing I could start again later, navigate all that knowledge in an infinity of different ways.

This impulse is why I am so good at reading the internet, and also why it's so easy for me to fall right into the internet's gaping maw at times of peak personal nervousness. The structure of the internet is perfect for the way I have wanted to read since I was a child: getting to the bottom of any question I ask, endlessly linking out to adjacent topics, the sense of horizonless knowledge at my disposal. But that structure isn't just useful for learning about R.E.M.'s IRS Records back catalogue. Trying to reach the end of the internet can be a harrowing quest. Reddit, Wikipedia, WebMD, even Twitter—these are black holes that can pull me in, turn my powers against me. The screen of my phone or the screen of my laptop can transform my readerly impulses into bizarre versions of themselves. Even things I love become grotesque in this space.

For instance, every now and then, I wake up with a headache that has an acute, obvious, and preventable cause: I stayed up until nearly 1 a.m. reading plot synopses of incredibly violent and narratively tortuous horror films on Wikipedia. If, in the pure light of the morning sun, you came to me and said, "Phil, you can *either* read one chapter of an acclaimed literary novel, have a good night of sleep, and wake up feeling refreshed and ready to greet the day, *or* you can read twenty-six plot summaries about people being flayed alive by secret cults, and then wake up feeling as if you have a mole family living inside your sinuses," I would clearly choose the former. Usually I do. But every once in a while, I'll get in bed, read four or five pages of a physical book, and then be struck by a thought in the very back of my mind: *I wonder what the filmmakers of the French New Extremity are up to?* Then I'll pick up my phone and read until my eyes are chapped.

The simplest way of explaining this is that, in ways I've already begun to detail, I am afraid. The most directly applicable fear here is that I am afraid of horror movies. Ghosts scare me, knife-related violence makes me want to barf, and demonic possession has always been a substantive, if outlier, concern of mine. But, like a lot of things that I'm afraid of, I also think about it a lot. Fear is, in this way, a species of interest. Excessive, curdled interest. When an eyeball gets sliced up or a ghost appears out of nowhere, it's not the turning away that's most important, it's the thrumming desire to return to the image. I'm *interested* in it, but indulging that interest by watching actual movies in this way is stressful and time consuming.

So I read about them. And apparently I'm not alone. Gabriella Paella, at *New York* magazine, who lurks around horror Wikipedia much as I do, wrote about the phenomenon in 2019—as did Rob Harvilla and Jaya Saxena. In the process of her investigation, Paella contacted a sociologist named Margee Kerr, who explained the nocturnal practice in this way: "Sometimes people need a little bit more of a protective frame. So instead of watching a horror movie, reading it can feel a lot less scary because it's more in the person's control."

This seems right, the way the screen of my phone—literally *in my hand*—restores the control that horror films are precision-built to steal from me, the way that it transforms terrifying images into terrifying concepts that I can more comfortably roll around in my head. Reading these plots reduces them to information, free of the visceral sting of the image. All the same, there's a kind of blunt shock value to the matter-of-fact wiki-style prose that evades the gothic stylization of the film genre itself. Reading these bare-bones descriptions can make a terrifying horror plot seem unfathomably dumb, but it can also make a story like that more unsettling.

This Wikipedia plot surfing, though, is deeply enmeshed for me within a network of other activities that have little at all to do with the cinema. Doomscrolling the news on Twitter, getting lost on NBA trade rumors message boards, mask shopping, armchair epidemiology, symptom checking, and, of course, reading about and watching videos of unidentified aerial phenomena—these are the spirits of the night. The positive contours of my daily screen time—watching film and TV series I love, texting and tweeting with far-flung friends, following my favorite basketball players, keeping myself and my family healthy, writing this book—these things are liable to undergo a kind of supernatural transformation on my phone when I'm otherwise stressed, when it is, as Ishmael puts it at the start of *Moby-Dick*, a "damp, drizzly November in my soul."

This screen time, I have learned, has a clinical name: revenge bedtime procrastination. This phenomenon is not exclusively about screen use. Originally just called "bedtime procrastination," the "revenge" was added in the American vernacular after a viral tweet from the journalist Daphne K. Lee in 2020 which revealed that there was already a term for this in China. The three requirements for a behavior to qualify as bedtime procrastination are that the person has to delay going to sleep long enough that it significantly reduces the total hours of sleep time, that this delay has to be completely unnecessary, and that the person has to be aware, to some degree or another, that this will have negative consequences. This cycle—staying up late for no reason even though you know it's bad for you—is related, by the psychologists and sleep doctors who study it, to the intensity of a person's daytime schedule. In the words of WebMD, one of my most frequent revenge bedtime procrastination tools, "it means you get 'revenge' for your busy daytime schedule by fitting in leisure time at the expense of shut-eye."

So bedtime procrastination could involve anything, but for me, and for a lot of people it seems, it involves screens. As I was writing this book, I asked the people of Twitter what idiosyncratic rabbit holes they fall into when they should be trying to sleep. The list is breathtaking in its breadth and in its specificity. I encountered numerous people who share my Wikipedia horror habit (plus a statistically significant subgrouping of people who read Wikipedia articles about famous serial killers); there were several people who watch instructional videos for musical instruments they do not play; several people watching explanatory videos about regional accents; lots of people doing consumer research for upcoming small or large purchases; people watching unboxing videos on YouTube; people falling into Reddit; people falling into MetaFilter; people falling into IMDb; lots of genealogical research into the royal family. One friend looks at the Facebook photos that her high school classmates post of their kids going to homecoming. One friend said simply, "moose YouTube." Everybody scrolling, nobody sleeping.

But it's not just internet spelunking. This is where binge-watching comes in, for instance, or at least the kind I used to succumb to. For the first two years we were in St. Louis, I, technically speaking, was *not* in St. Louis. We'd both gotten great jobs, but they were at disparate ports of call on the Mississippi River, so I worked in Baton Rouge, Louisiana, but I quote-unquote lived in STL. We were advised to do this. As commutes like these are common for academic couples, lots of people had words of wisdom. One of those was that, for the health of your identities and relationship, you should select one of the cities to be "home." Keep all your stuff there, have your pets live there, rent the two-person apartment there. In the other location, live like a single grad student again. This way, rather than having the feeling of leading separate lives, one of us would convince ourselves to have the feeling of merely being on a never-ending work trip. We both liked STL more anyway, so I rented a room from a colleague at LSU, and I lived like Kato Kaelin for two years.

It was lovely in lots of ways, but also lonely and considerably more stressful than I would or could allow myself to acknowledge day-to-day. I would work all day, sequestered in my sweaty office, endeavoring to "write my way" into living with my partner again, and then I'd leave campus, drive back to my room by way of Raising Cane's Chicken Fingers, and eat dinner on my bed and watch TV series on my laptop.

The particular flavor of screen time I most hankered for in those moments was comfort food. So I'd watch *The Office* or *Parks and Recreation* or *Deadwood*. I'd watch *Mad Men*. Series dense with internal detail for me to discover and rediscover; series with the kind of emotional or nostalgic density that could bring me comfort. Series I'd watched already, watched first, with Mel. Shows to play with: what if I rewatch *The Office* with the idea that Dwight is the protagonist? What if Pam's the protagonist? What if it's a tragedy about a man whose life is ruined by reality TV? What if Creed is a ghost? Episode after episode. The screen would ask if I was still watching. I would say that I was. Chicken finger after chicken finger. Often, I'd wake up at two or three in the morning to realize that I'd fallen asleep watching and four or five episodes had gone by as I slumbered in my work clothes. This sort of revenge wasn't necessarily about how busy my day was. It was about searching in vain for a feeling that'd been missing from my day. The feeling of home.

* * *

So the internet can be a tool for turning even my hobbies, even the things I love, into monstrosities, and bedtime is its prime time. One thing I love, for example, is basketball. I've loved it since I was a kid in the screen time nineties, but I never had a pro team to watch. I rooted for Michael Jordan's Bulls, and I wanted Charles Barkley to win with the Suns, and I collected hats from all the teams whose logos I loved, but, growing up in Pittsburgh, I didn't have what you could call a rooting interest. And if we wanted to see a pro basketball game, we drove to Cleveland. The Cavs weren't always bad during those years, but when we'd get tickets to see whatever team of superstars were on the road in Cleveland, it felt like getting tickets to a Harlem Globetrotters game at the Washington Generals' home arena.

So while it might have been a logical—even a virtuous—thing to become a Cavs fan, I didn't. That's a failing on my part. But as the years have worn on, and I haven't lived in Western Pennsylvania in a while, I've grown affectionate of Cleveland from afar. So it was with warmth, and a genuine rooting interest, that I watched game 7 of the 2016 NBA Finals—between LeBron James's Cleveland Cavaliers and Steph Curry's Golden State Warriors—in my parents' condo in downtown Pittsburgh with my mom and dad and Mel, and an infant Maeve asleep in the other room. Since I'd lived away from home, I'd managed to see the NBA Finals with my parents enough that it had almost become a tradition. It's in June, so I was often just home from college, and since it's generally around my dad's birthday as well, it would often coincide with visits after I no longer went home for the whole summer.

But that year was different. Two nights after we watched Cleveland win the championship, Mel and Maeve and I drove back home, six hundred odd miles to St. Louis. And two nights after that, my parents, who have lived in Pittsburgh their entire lives, took all their stuff, left Pittsburgh, and moved to St. Louis with us to help us care for Maeve.

LeBron James has an unusual relationship with his hometown. Born in Akron, drafted out of high school by the Cavaliers, he tried for many years to

Figure 4.2. Game 7, NBA Finals, 2016

win in The Land, as they call it, but failed. So he left in free agency to join his friends and win a handful of titles in Miami. When he decided to come back to Cleveland after his quest in South Beach, he said that his time down there was like going away for college. He grew up a lot, and he learned how to win, etc., and he decided to bring that back home to Cleveland. I'd love to be able to draw at least this one comparison between myself and King James, but I can't. When I went away to college, I didn't come back home. I went to two grad schools, and I got jobs in far-flung locations, and I made new homes in some of them. (I'd be interested to know what LeBron's grad-school option would have been.)

When LeBron won that game 7 back home, though, we were with him. Iconic moments flowed from the screen; we were all rapt. There was The Shot and The Block and The Weird Moment Where It Looked Like He Broke His Hand Or Something.

And when he collapsed crying on the ground at the end, I cried, too.

I was really happy to be able to be there to see it in Pittsburgh, with my mom and dad. I cried because LeBron was so happy, because it was such a dramatic game of basketball, because you could see him work every second of that game to keep this one thing for Cleveland, because when he collapsed you could feel the weight of all of that guilt and all of that hope, because my parents were leaving their home for me and my family, because it felt like I was leaving Pittsburgh again too, because I thought about my daughter being eventually old enough to stay awake to watch the NBA Finals with us, because I could feel that we were all in the same place and sharing the same thing, because LeBron left Cleveland but he always loved Cleveland, because I left Pittsburgh but I've always loved Pittsburgh, because I don't know LeBron and I am not LeBron but I think I understand at least one thing about him.

Pittsburgh didn't win the NBA Finals, but that game was somehow about me and my parents and our city. As long as I live, I will never forget watching game 7 with my family in that place, witnessing the spectacle of a person doing something that beautiful and that hard for the people who believe in him.

But that's not the only kind of basketball screen time. Every year, there's a very special day called the NBA trade deadline. It's a day in the middle of the professional basketball season that's the last day players can be traded to other teams before the playoffs. (Free agency day is the day in the offseason when players choose where they go; trade deadline day is the day when teams choose.) Rumors are always swirling about possible trades, possible broken relationships between star players, possible ultimatums given from stars to team owners, but they hit a crescendo the week leading up to the trade deadline.

The deadline is fun, when it's fun, because it gives fans the opportunity to reimagine their teams, to see that their teams have enough faith in their players to go get reinforcements, or to see that their teams are loading up young players and draft picks for future glory. It's a day about possibility, as much as it's also often about bad feeling.

But it *is* also about bad feeling. Players don't always want to be traded, and, beyond that, focusing on trades as entertainment can be crass at best and dehumanizing at worst. This is the time of the year when sportswriters tend to start referring to players as "assets." Players in the contemporary game have a lot of agency, but the trade deadline is a day that celebrates their temporary loss of agency. Or, rather, it's a day for managers, for owners, for the business offices that employ the players we love, men traded for other men, people reduced to their abstract value.

Regardless, it's a huge day for sports media, and thus a huge day for screen time for me. There's a multitiered architecture of reporters who track trade rumors, and they're all online, posting their tips and sourced rumors as currency. These deals are complicated, often executed with minutes to spare, and they are conducted in super-secret. Even among clued-in journalists, very little about the dynamics of these deals are ever fully known.

So reporters, and fans, become detectives. At the top are a few highly paid "NBA Insiders," specifically ESPN's Adrian Wojnarowski and *The Athletic*'s Shams Charania. When you see these men on television in deadline week, if they're not being directly spoken to, they're looking at their text messages like oversaturated teenagers. They text constantly. They check email constantly. They are, to me, the affectless gods of screen time because their only skills are that (a) they know lots of people who will text them information that they want to be tweeted out and (b) they can tweet out the information they received very quickly. When Wojnarowski breaks news on Twitter, it's called a "Woj Bomb." Recently, Shams was asked about his screen time in a profile in the *New York Post*. He said that, on average, he logs seventeen to eighteen hours per day, but that, during the trade deadline, it often hits twenty.

I can't match that energy, but my app tells me that I have an uptick during deadline season, too. And I'm not just watching Woj and Shams. There are other tiers. Below those gods, we find national reporters and analysts who are around teams enough to hear things. They write articles about rumors, and they tweet, and they speculate about the veracity of other peoples' tweets. So do the beat reporters for national and local services, and so do various local insiders in specific team markets.

Then it gets weird. There are anonymous posters on Reddit who claim privileged information. There are posters on basketball forums—like RealGM—who post elaborate, impossible, fantasy trade concepts. Sometimes there are people who actually know what they're talking about, hiding in plain sight on the forums. They tend to disappear when they attract too much attention. And then there are the fakes: Twitter accounts claiming to be insiders, producing realistic facsimiles of privileged information.

I read all of these, and I read them with abandon during deadline week. I'll dig through message boards for my team—since meeting Mel, I've adopted the Sixers—as well as the rest of the league. I'll send dozens of round-the-clock DMs to my friend, and fellow NBA rumors hawk, Jorge. I'll refresh, refresh, refresh Twitter in the hopes that one day I'll see the Woj Bomb first, that victorious "3s" next to Woj's name on the timeline. My "pickups" from my pocket skyrocket on deadline day, like a gunslinger in the Wild West. And my bedtime procrastination sinks to new depths.

This past deadline day, I went to bed around ten and started reading a novel by the Mexican novelist Fernanda Melchior, but then I thought: *I wonder if I can prove if Naismith2Silver is real.*

Naismith2Silver is a Twitter account with eighty-one followers, myself among them. Its avatar is a black-and-white photograph of James Naismith, the inventor of the game of basketball. (The "Silver" refers to Adam Silver, the NBA's current commissioner.) It posts very rarely, but when it does, it only posts elaborate backroom narratives of NBA trade negotiations, drawn, the poster claims, from a single, high-level league source. Every tweet from Naismith2Silver is allegedly a direct quotation from this source that the poster copies and pastes from their text message chain. The poster includes their own questions, in the interest of transparency. A long thread of quoted texts about a possible Ben Simmons for Russell Westbrook trade might begin with, "And what about the Lakers? Are they getting involved?" The poster adds hashtags to all player names, I guess in the hopes of achieving virality. As I've said to Jorge, Naismith2Silver is *either* a person who knows a very highly placed NBA source or Naismith2Silver is a spam bot. Either one of those eventualities is wholly plausible to me.

That night, I decided to see if I could figure it out. I scrolled back years into the account, getting timestamps for seemingly correct predictions about trades or draft picks or trade requests, and then scrolling years back into Woj and Shams, trying to find initial timestamps for this information, find the news stories where they originally published, to tell if Naismith2Silver had just quickly produced a fake that would seem real in the scrum of the moment or if they'd really been ahead of the game.

I scrolled and scrolled until I started getting a searing pain over my right eye. This is a pain I've often felt in moments like these, where the scroll gets out of my control, when I feel at the whim of my screen rather than the other way around. I looked up and saw that it was 2 a.m. I hadn't proven whether Naismith2Silver was real, but it wouldn't have mattered. Who even would I have been proving it to? I'd fallen in.

I turned off my light, that night, with shame and a sick feeling. A thing I love—basketball—that brings me joy, had become something else. I remember watching the Sixers' first playoff wins of the Joel Embiid era alongside Mel on our TV in our living room; I remember seeing a cursed arena employee accidentally fire off a confetti cannon when the Sixers only tied, but did not win, a playoff game on a TV in a hotel in Milwaukee; I remember all the Jordan series on the RCA Colortrak; I remember watching, with my parents on the TV in their bedroom, Chris Webber call a time-out and lose the NCAA championship; I remember watching an inspirational movie about Pistol Pete Maravich with my dad when he was trying to convince me to practice more in grade school. And I remember LeBron crying on the floor with me and all the people I love in the room together. What goes on between me and Naismith-2Silver at 2 a.m. is not that. Mel's asleep, Naismith-

2Silver is nobody and nowhere; it's just me and the screen, alone together. I went back to my novel the next night chastened, but relieved. My eyes didn't hurt anymore.

Revenge bedtime procrastination online is a strange, messy activity. It satisfies an impulse, but it isn't necessarily pleasurable. As cultural critic Anne Helen Petersen clarifies, after she's spent forty-five minutes scrolling in bed, "I haven't been having *fun*." In that sense, it's a kind of symptomatic reflex of the screen time consciousness, a rote, joyless kind of attention. It takes its shape from the totalizing, encyclopedic networks of the internet, but it's largely without content, discontented, even. Screens provide access, and they provide information, but they can't, on their own, provide presence. Within this network, we can go anywhere, and we can learn anything, but we cannot, or will not, find rest.

5

SPACE JUNK

The part that most worried me about our road trip in the summer of 2021 was, ironically, the part where we would be least likely to catch the novel coronavirus. After traveling to Pittsburgh, to the Jersey shore, and to Philadelphia, with a brief pit stop in Kentucky's Red River Gorge, we would head homeward to St. Louis, but we would do so by way of a long weekend in the Blue Mountains of Virginia. Mel had scoured the internet to find a place for us to unwind after a busy, several weeks–long trip, and she found an idyllic farmhouse, nestled alongside a babbling creek, in a remote valley miles and miles off the highway. We invited my best friend, Thom, newly married, and his partner, Emily—who, for pandemic reasons, we had not even met yet—to join us from Philly. The girls would go creek walking, we would grill and talk and laugh and play board games, and we could live, for a few days, in a kind of quiet that was inaccessible to us in our ordinary lives. We could be together. What worried me most was that we wouldn't have cellular reception.

Looking back, I'm not sure precisely what made me so afraid of this. As I've detailed at length in this book so far, my phone is where a lot of my anxieties live, or, at the very least, it's the medium through which I feel them. Having a few days away from screen time—which, for me, had been filled with ominous portents about new viral waves and ominous portents about unidentified aerial phenomena all that summer—seemed like it should come as a relief. But the idea of a screen-free weekend was filling me with a creeping dread.

Being "out of reach," in ways both vague and specific to me, was the central worry. I am lucky enough, for instance, to have my parents with us in St. Louis, but, as we headed east to close the distance between the girls and their faraway family—my grandpa and aunts and uncles in Pittsburgh and Mel's parents and brother and sister and aunts and uncles and cousins in Philadelphia—I returned to the now mercifully rare state of feeling distant from my *own* parents. I worried about the fragility of our luck and good fortune, the possibility of losing it if I let go—virtually—for even a moment.

And it only started with them. What if somebody got hurt at the farmhouse, and we had to figure out how to get them to the hospital? What if this band of college professors and pro-vax healthcare professionals—Thom and Emily are nurses—drinking our craft beers and eating our fake hot dogs were beset by any one of the numerous locals whose houses, festooned with MAGA and AR-15 and "Fuck Fauci" flags, we'd driven past on our way to our cute little farmhouse? I've read several Wikipedia plot synopses for films in which this exact thing occurs! What if some world calamity transpires while we're safely ensconced in our screen-free bubble, and the world we emerge back into is one devasted by war and death, or, at least, more devasted than it had been when we entered the bubble? What if the asteroid finally hits, and we don't even know about it until the shockwave comes?

These reasons all help to explain why the Virginia trip gave me the creeps, but I came to realize, beneath all this, there was an additional reason. And that reason was harrowing in its literalness: I didn't want to be without my screen. I didn't want to experience what it would feel like to pick it up—178 times per day—and have it not look back at me. I have a relationship to that screen, and I didn't want it to pause. I have a *self* with that screen—an imperfect self, an anxious one, but a responsible and meaningful self too. Who would I be without it?

It's times like these when I feel most swayed by all the screen time prophets who talk about it as an addiction or a pollutant or a toxin. It's when I feel most convinced that screen time *should* make me feel bad. My consciousness, my body, shouldn't be so reliant on this toasty little machine. It shouldn't be this stressful to give it up.

Mel could sense my anxiety about this part of the trip. She's familiar with this dynamic, and she knows I'm already pretty deep in if I'm willing to acknowledge it out loud to her. I'd already let her know about my concerns regarding the various, mysterious faster-than-light craft zipping up and down the Eastern seaboard in June. She helpfully reminded me that I didn't need to be losing sleep over black triangles, and that I was just worried about us catching COVID on our trip. That was both true and useful to hear. But she could sense my anxiety about Virginia, too. In an act of almost surreal generosity, during the second week of the trip, she even offered to let us just drive straight home and skip this dogleg that she'd been so looking forward to. Accepting that offer would have felt like a kind of escalation that would have sent my day-to-day anxieties into a wild downward spiral. They would have gone from recreational to real. I would not have returned home free of anxiety. I would have felt guilt and shame that I'd let this screen build an obstacle in my actual life, that this silliness would affect other people, people I loved. That she even suggested it was a bit of a shock. *Is it that bad? Is this what her love for me demands that she do? Is this creature made of screens so completely who I am?*

So we went. The journey was an absolute nightmare. We went up and down and up and down steep mountains for seemingly hours, the girls puking their guts out in the backseat from car sickness, Mel and I holding in our own barf from the incredibly noxious smell of regurgitated Wendy's filling the cabin of our Mazda. But then we arrived. The valley was calm. The house was cozy. We could walk from the back deck for five minutes and wade into a gleaming creek filled with tadpoles and little fishes. We even had lovely neighbors raising friendly bull mastiffs in a large enclosure next to our yard. More than anything, it was the sight of these monstrous dogs, who looked like what Rick Moranis turns into at the end of *Ghostbusters*, placidly whiling away the days that filled me with tranquility. They trotted alongside us as we walked to the creek. They asked us to pet them. They were scary, but we were not scared.

And we could look up. The planes that flew overhead were not even close to their landing patterns, as the planes that regularly buzz our house in South St. Louis all are. I knew from the spottiness of our car's GPS navigation system on the way that even the satellites could barely see us. We could come outside at night and see millions of stars, even some of the celestial clouds that surround them. There was no light pollution from our cities, no sounds of motorbikes revving past, no signal, from AT&T or from the UAPs.

It's a well-known fact that iPhones are capable of taking only deeply shitty pictures of the Moon. Twitter and Instagram are chockablock with such pictures whenever there's a Blood Moon or a Harvest Moon or a Sturgeon Moon or whatever. People see these rare heavenly sights, and it's natural—or it's become naturalized—to want to capture the moment. But the phone in my pocket can't do it. It's a good metaphor because it's such a reliable phenomenon. Your phone can't take a picture of the Moon, certainly not the stars. You can't have them—they don't belong to you. Every photo shared of these celestial bodies is evidence of all that a screen can't do, can't capture, can't open up for us.

I didn't try to take any photos of the stars while we were in Virginia. I didn't feel small gazing up at the night sky, nor did I feel a need to control the situation in the way my screens so frequently enable to me try. I felt together with the love of my life, with my best friend, even with my sleeping girls in their beds inside. I felt like I was a part of an infinite universe, or, at least, that I was at home on this planet with all my friends and loved ones and everyone who'd ever been or would be. In one of Maeve and Phoebe's favorite picture books, the Northern Irish artist Oliver Jeffers reassures his infant child, "You're never alone on Earth." My phone, and its limits, would have said otherwise. My phone would have told me what it couldn't see. Or it would have showed me a picture that captured nothing of what I felt in that moment. It would have been a misrepresentation. It would have flattened a time spiked with love and memory and fear and relief and awe. It wouldn't have been able to tell me the truth about what I was looking at or what it meant. Had I picked it up in that moment, it would have told me a lie.

Eventually, though, we had to leave Virginia, drive through some hills until the satellites saw us again

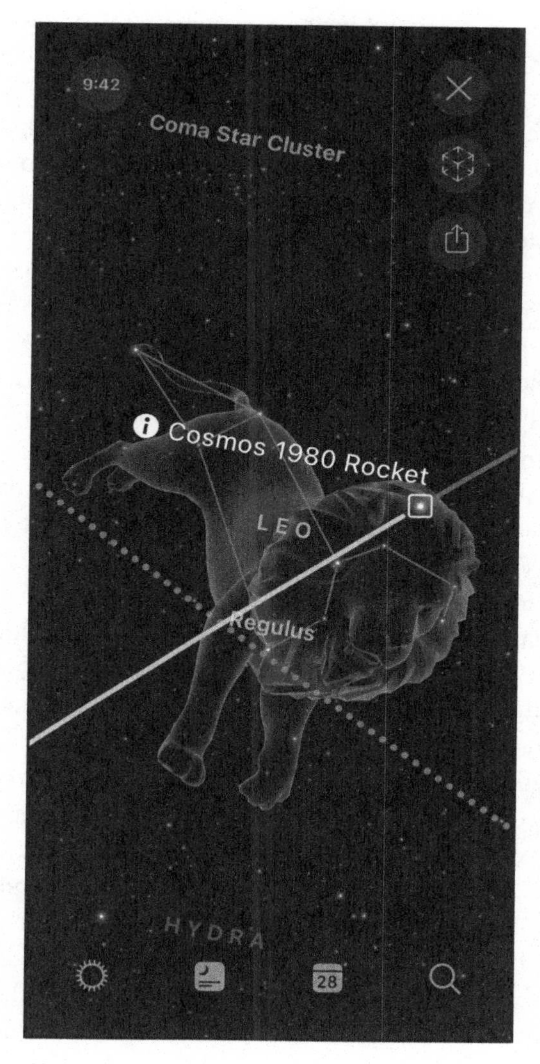

Figure 5.1. Sky Guide, somewhere in South St. Louis

and they could guide our car home to the Midwest. That short screen time break didn't necessarily exorcise my demons, but it did serve as a bit of a reset. When we hit a main highway, I pulled out my phone to see if I'd gotten any texts, to check Twitter to see if any fun trades had gone down at the NBA draft (which had occurred that weekend), to make sure no asteroids hit, to text my parents to say we were safe and on our way. But I put it back down again, thinking of all the things I'd seen without its help.

Before we left for the trip, in order to help proactively assuage my building anxiety about the Virginia stopover, I'd downloaded an app called Sky Guide to my phone. I don't think, at that point, that I'd fully understood precisely *how* offline we'd be in that valley, and I got the app thinking it'd be fun to use when the stars came out. What Sky Guide does is provide a map of the night sky, enabled by your phone's location data. You point it wherever you're looking, and it identifies stars and planets as well as constellations and whatever else is up there. You can click on anything to read wiki-style explainers, and you can search the app, too. If you're looking for Betelgeuse but can't quite spot its location, you can type it in, and Sky Guide will swirl its point of view around to find it. You can look, through this app, at things you can't see, knowing that their invisibility is the result of your human eyes and the world that humans have built. It shows you what you would see if you could see everything.

Maeve loves it—she calls it "star app." She's rarely up late enough for us to actually stargaze with the app in our backyard—though I do, usually when I'm locking up our back gate or making a trash run to the dumpster in our alley. But she likes to use it in the slim half hour before she goes to bed, between her shower and her toothbrushing and her outfit planning and whatever chapter book about twelve-year-olds we're reading at the moment. In her room on the top floor of our house, she whips my phone around looking for her favorite constellations, searches for planets she likes, reads about Capricornus and Aquarius and Alpha Centauri. It's a pretty striking sight: a bedtime six-year-old with long wet hair in her pajamas at 7 p.m. central time in a room with all the blackout curtains shut, rolling around and stargazing.

But she's seeing distant stars just about as much as I am when I go outside with my Simple Human trash bag in my hand, gazing up at the handful of celestial orbs I can make out through the light-polluted sky of my St. Louis city backyard. It is, to her, magical. It's the kind of in-depth introduction to the fathomlessness of the universe that would have simply been impossible for me to have at her age. And it demands her time and her attention. The screen shows her what she could see if she could see everything, but, more than that, it tells her that, at whatever scale she imagines the universe, she should think bigger. The brilliance of Sky Guide is that, while it uses an interface that resembles Wikipedia—and reminds me of my various, revenge bedtime quests to reach the end of the internet—it's well aware of its limits. Maeve will never reach the end of outer space. If she ever reaches the end of Sky Guide it'll simply be because there are so many things even the app doesn't know.

This is screen time, too. I don't know what it means, whether it's a contraction of her imagination or an expansion of it, but when, night after night, my daughter pleads to me to give her my phone to explore the stars in the sky, I hand it over. And it doesn't feel bad.

* * *

My favorite part about Sky Guide isn't the stars and the planets, necessarily: it's the space junk. In addition to all the organic celestial bodies it maps, Sky Guide also helpfully keeps track of the man-made detritus floating around in orbit. It's mostly rocket husks, a quick glimpse of the International Space Station if you're lucky, and some of it is even visible to the naked eye. Seeing a shooting star is always a small miracle, but, almost by definition, Sky Guide doesn't map those sorts of contingencies. What it *can* map is what it knows is up there, and we know that there's a lot of junk in space. There's so much junk, in fact, that scientists and astronomers are worried about it—as of 2021, there were twenty-three thousand pieces of man-made debris larger than a softball orbiting the planet. The major concern is collisions.

Even a tiny piece of debris can meaningfully damage a spacecraft, or, at worst, trigger something called the Kessler effect, where an isolated collision ends up destroying dozens of satellites that perform essential, terrestrial functions in rolling, exponentially more disastrous waves.

The more existential concern is that all this junk—even and including working satellites—is going to impact our vision. The most amazing thing about identifying space junk with Sky Guide is that, sometimes, when I look, I can actually see it up there. A small dot, moving slowly, but with purpose, through the dark blanket of the sky. But this momentary charge of excitement is a symptom of a broader problem. According to scientists, the sheer volume of space junk—much of it visible to the guided eye—has increased the brightness of the night sky by as much as 10 percent. Everything from rocket husks to satellite constellations designed by private companies like SpaceX can ruin astronomical observations by scattering sunlight or photobombing long-exposure images of distant objects. The more we put up there, the less we might see.

One of those man-made objects that my app might one day see is the James Webb Space Telescope. Out beyond the ring of space junk that encircles our world, this telescope has fixed its gaze on the farthest reaches of the universe. As I finish writing this book, we've only now just begun to see the highest-resolution images ever produced of deep space. And they're not just images of planets and galaxies and nebulae. What we're seeing is a picture of time. The Webb telescope can see far enough that it can look into the distant past, thirteen billion years, almost the beginning of the universe itself. My Sky Guide app might find the Webb telescope and tell me where it is in my, real, time. But what that telescope is seeing and taking pictures of is something that feels outside of that real time. The James Webb telescope's screen time is taking place in unreal time.

Anticipating the first pictures, these graphic records of intergalactic time travel, I was not expecting alien craft or suspicious shapes. I was expecting impressionistic daubs of color and light. I did not anticipate them, in other words, with fear. But, perhaps because of Sky Guide, perhaps because of last summer's encounter with the clean night sky over Virginia, perhaps because I'm just now finishing a book about screen time and its capacity to bring us together, I was more interested than I otherwise might have been.

My idea was that we would all look at the first image together, when it was first released, that we could all sit there like Don and Peggy watching the Moon landing, to witness this epochal screen time moment. The stars as we've never seen them! The stars *when* we've never seen them! Our place in the universe revealed at last, a sky unfathomable even to Maeve's app. A moment of connection, a family.

Unfortunately, the Biden administration was apparently not aware of how much I'd hyped up this screen time experience. Their big, live reveal of the first image—which wasn't even covered live on cable news—was a bit of a dud. Instead of evoking the mysteries of the cosmos, the livestream mostly evoked the now all-too-common experience of a Zoom meeting run by somebody who doesn't know how to screen share. When the image finally appeared on my laptop, Mel and I spoke rapturously to Maeve about it and its implications, but our reveries couldn't overcome the bungled rollout. She was unimpressed. Phoebe, sitting on Mel's lap, chimed in: "I want to watch *Bluey*." So we did.

Later that night, as Maeve readied herself for bed, I showed her the image again—this time a high-resolution one I'd downloaded from the NASA website. "I don't really want to talk about that," she said. And I get it. Despite all the hullabaloo, it's an image I have a hard time looking at myself. If I stare at it with the ironized gaze of a Twitter user, I'm scanning to think of images it reminds me of: movie theater carpeting, the final pages of Jon Klassen's picture book *We Found a Hat*, the last scene of Claire Denis's *Beau Travail*, the cover of the Pavement album *Terror Twilight*, a Trapper Keeper. These jokes give me distance from an image that's undoubtedly beautiful but undoubtedly upsetting, too. If I look too hard at it, I realize I no longer have any capacity to understand it. It's too far out of space, too far out of time.

Its understanding of me is so much more complete and complex than my understanding of it, or even my understanding of myself.

But there's also the fact that even this image, like those iPhone pictures of the Moon, is not quite telling the truth. The image that NASA released is in glorious color, but those colors are merely indices of various environmental factors and light-emitting gases. This image isn't, to put it simply, what any of that would have looked like had our human eyeballs been there to see it. This is a long tradition. When the Hubble telescope was sent up, NASA developed what's called the "Hubble palette," which is an image processing technique that assigns different colors to different types of data in an image in order to render visible features that would otherwise be invisible to the naked eye. The colors are not, technically, natural. In fact, the art historian Elizabeth Kessler has written about the way that the Hubble palette is as much influenced by nineteenth-century aesthetics of the sublime as it is any verisimilar recreation of the stars above. The Webb images, likewise, are as much data visualizations as photographs.

Sky Guide lacks the grandeur of the James Webb telescope and the Hubble palette, but it's trying to do the same thing: bring something distant and unimaginable a little bit closer to us. After rejecting the Webb image, Maeve said simply: "star app, please." I handed her my phone, and I watched her reel around her room, methodically locating, or running searches

Figure 5.2. Maeve Anna Maciak-Micir, July 28, 2022

for, each planet in the solar system in order. She was immersed, she was connected, she was thrilled to be part of this neighborhood in this universe and giddy at the idea that the phone could tell her where, precisely, all her neighbors were at the moment.

I didn't feel small, watching her immersed in her screen which was immersed in the sky over Australia where Neptune would have been visible that night. I didn't feel alone. The universe is very old, but we're not. There'll be more images, and when they arrive, Maeve will be older, and so will I. Sky Guide will have helped her get her footing in this galaxy a little bit, just as the Webb pictures will unsettle that footing the more she learns about them. But there'll be other telescopes taking pictures of the same planets and stars and galaxies, and we'll see them where we are when they arrive. And, in between, we'll look at our phones and our laptops and our TV sets, and we'll watch *Mad Men* and *Bluey* and the Super Bowl and the faces of our loved ones appear and disappear. And all that will happen in screen time, which is a kind of time, a constraint. This is what it's like. This is how it feels. This—the charged space between these people, their hot screens, and the stars exploding into imaginary swirls of color at the very beginning of the universe—is what's happening.

Acknowledgments

I wanted to write this book specifically in order to write it with my friend Sarah Mesle. *Screen Time* represents, for me, a pretty big step away from a former writing life into a new writing life, and it's a step I only wanted to take with someone I completely trusted. I've known Sarah, and I've been writing with Sarah, for a long time, and I knew that, with her eye on things as editor, protecting what needed to be protected, coaxing me away from my worst habits, I'd be able to write a book that sounded like me. I want to thank her for that, and I want to thank her and Sarah Blackwood for founding *Avidly* in the first place, for being my intellectual heroes, and for giving me the courage to wear a jean jacket in professional settings.

I consider this book to be one really long entry into the *Dear Television* extended universe, so I want to thank Aaron Bady, Jane Hu, Evan Kindley, Lili Loofbourow, Sarah Mesle (again), and Anne Helen Petersen for being the addressees of yet another overlong, sentimental essay from me. I started writing with these people ten years ago, and it was through writing with them that I realized the kind of writer that I wanted to be. I'm so lucky to have them as my friends and readers. I write to them, always.

Thank you to my editors at NYU, Eric Zinner and Furqan Sayeed, for being onboard with all of this and for supporting a series that gives writers this kind of creative space, and to Ainee Jeong, Dan Geist, and Jenny Rossberg for taking such care. And thank you, as well, to the editors I've worked with on other projects during the writing of this book—Sam Adams, Lisa Borst, Jane Hu (again!), Laura Marsh—whose off-topic insights eventually made their way in here. Thanks to my colleagues at *LARB*, specifically Annie Berke, Boris Dralyuk, and Lauren Kinney Clark. And thanks to Akin Akinwumi for his excitement, support, and *time*, as well as fellow Pittsburgher Ed Simon, for the assist.

A few years ago, I wrote a piece for *Slate* about the legacy of Arnold Lobel's *Frog and Toad* books among contemporary children's authors. In ways I've come to realize since then, the conversations I had for that essay led pretty directly to the idea for this book. So I want to thank Marissa Martinelli for accepting that pitch at the beginning of the pandemic, and I want to thank, especially, Mac Barnett and Jon Klassen and Kyo Maclear for talking to me about what it means to see the world through the eyes of children.

This book is, in some ways, my ode to television studies. And I want to thank some of the scholars I've read or corresponded with or who've worked with me at *LARB*—many of whom I've met at some point or other—all of whom have given me an education in a field I love: AJ Christian, Racquel Gates, Elana Levine, Myles McNutt, Jason Mittell, Michael Newman, Nick Salvato, Lynn Spigel, and especially Kristen Warner, who, if she's ever been skeptical of my investment in this stuff, has hidden it incredibly well. I'd also like to thank all of the working critics who've been supportive, interested, or at least fleetingly engaged with the work I've been doing at the margins of their profession for all these years. In particular: Angelica Jade Bastién, Roxana Hadadi, Jack Hamilton, Alison Herman, Lauren Michele Jackson, Emily Nussbaum, James Poniewozik, Soraya Roberts, Matt Zoller Seitz, Alan Sepinwall, Philippa Snow, and Kathryn VanArendonk. Thanks to Jorge Cotte for being one of my favorite writers as well as the recipient of my most deranged basketball thoughts. And thanks to Dave Alff, Dave Goldfarb, and Lindsay Reckson, who might not realize it but who showed me a really meaningful amount of support when I started writing about being a parent.

Thanks to all my students at Washington University in St. Louis, especially those of you who've helped me build the "Hot Takes: Cultural Criticism in the Digital Age" seminar in American Culture Studies. Talking with all of those brilliant, inventive students about what criticism is for has helped me, well, figure out what I think criticism is for. Thanks also to my colleagues at WashU and in St. Louis generally: Pannill Camp, Noah Cohan, Danielle Dutton, Heidi Aronson Kolk, Ted Mathys, Paige McGinley, Heather McPherson, Edward McPherson, Marty Riker, Ignacio Sánchez Prado, Vince Sherry, Karen Skinner, Rachel Greenwald Smith, Abram Van Engen, Rebecca Wanzo, and Rhaisa Williams. It's good to feel at home here. Thanks also to Erin and Bobby Herrera, for being radical screen time role models.

Thank you to my parents, Barry and Sandy Maciak, for reading everything I've ever written, from my first essays on the Apple II to the present; for watching *DuckTales* and *The West Wing* and the NBA Finals and *The Big Chill*—over and over—with me; and for the active, loving part they've played every day in my life and that they now play in the lives of my daughters. Thank you for always being there, and here. And thanks to Thom Matera, for always being here, and there.

This is a book I wrote about watching TV with my kids and how that's changed the way I see everything else. A lot more about my life has changed because of them however, and, for that, I want to thank Maeve Anna and Phoebe Jean. Your attention and attentiveness to this weird world is miraculous to me. I love you both very very much.

And Mel. In 2011, Mel, who had been listening when I told her about how I'd always wanted to be a critic, told me, very simply, that I should do that. And not be afraid. Everything I've written since then is because of her, (occasionally very lightly adapted from conversations I've had with her), and dedicated to her. I can't thank her enough for that gift, and I hope, in time, to repay it.

Works Consulted

Bady, Aaron. "The *WandaVision* Cul-de-Sac." *Los Angeles Review of Books*. March 18, 2021. https://lareviewofbooks.org.

Blackwood, Sarah, and Sarah Mesle. "This Week in Ferrante." *Avidly*. March 25, 2016. https://avidly.lareviewofbooks.org.

Christian, Aymar Jean. *Open TV: Innovation beyond Hollywood and the Rise of Web Television*. New York: New York University Press, 2018.

Crawley, Ashon. "Do It for the Vine." *Avidly*. August 14, 2014. https://avidly.lareviewofbooks.org.

De Groot, Michell, and Annie Berke. "To Suffer a Witch in *WandaVision*." *Public Books*. November 1, 2021. www.publicbooks.org.

Denson, Shane. *Discorrelated Images*. Durham, NC: Duke University Press, 2020.

Engelhardt, Tom. "The Primal Screen." *Mother Jones*. June/July 1991.

Feuer, Jane. *Seeing through the Eighties: Television and Reaganism*. Durham, NC: Duke University Press, 1995.

Foucault, Michel. *The Order of Things: An Archeology of the Human Sciences*. New York: Vintage, 1994.

Goethals, Gregor. *The TV Ritual: Worship at the Video Altar*. Boston: Beacon Press, 1981.

Gordon, Jeremy. "Vine, the Only Good App, Is Shutting Down." *SPIN.com*. October 27, 2016. www.spin.com.

Guernsey, Lisa. *Into the Minds of Babes: How Screen Time Affects Children from Birth to Age Five*. New York: Basic Books, 2007.

Heitner, Devorah. *Screenwise: Helping Kids Thrive (and Survive) in Their Digital World*. New York: Routledge, 2016.

Herman, Alison. "Previously On: How Recaps Changed the Way We Watch Television." *Ringer*. July 31, 2018. www.theringer.com.

Holdsworth, Amy. *On Living with Television*. Durham, NC: Duke University Press, 2021.

Hu, Jane. "GIF Typologies and the Heritage of the Moving Image." *Hyperallergic*. September 28, 2012. www.hyperallergic.com

Hughes, Jazmine. "Vine Dries Up. Black Humor Loses a Home." *New York Times*. November 1, 2016.

Jackson, Lauren Michele. *White Negroes: When Cornrows Were in Vogue . . . and Other Thoughts on Cultural Appropriation*. Boston: Beacon Press, 2019.

Kamenatz, Anya. *The Art of Screen Time: How Your Family Can Balance Digital Media and Real Life*. New York: PublicAffairs, 2018.

Kessler, Elizabeth. *Picturing the Cosmos: Hubble Space Telescope Images and the Astronomical Sublime*. Minneapolis: University of Minnesota Press, 2012.

Klein, Amanda Ann. *Millennials Killed the Video Star: MTV's Transition to Reality Programming*. Durham, NC: Duke University Press, 2021.

Lange, Alexandra. *Meet Me by the Fountain: An Inside History of the Mall*. New York: Bloomsbury Publishing, 2022.

Lewis-Kraus, Gideon. "The U.F.O. Papers." *New Yorker*. May 10, 2021.

Livingstone, Sonia. "The Rise and Fall of Screen Time," in *Masters of Media: Controversies and Solutions*, ed. Victor C. Strasburger. Lanham, MD: Rowman & Littlefield, 2021.

Lotz, Amanda. *Cable Guys: Television and Masculinities in the Twenty-First Century*. New York: New York University Press, 2007.

———. *The Television Will Be Revolutionized*. New York: New York University Press, 2007.

Martin, Brett. *Difficult Men: Behind the Scenes of a Creative Revolution: From* The Sopranos *and* The Wire *to* Mad Men *and* Breaking Bad. New York: Penguin, 2013.

McNutt, Myles. "Game of Thrones—You Win or You Die." *Cultural Learnings*. May 29, 2011. https://cultural-learnings.com.

———. "Recap by Default: Why Terminology Matters to How We Write about TV." *Cultural Learnings*. August 16, 2013. https://cultural-learnings.com.

Mittell, Jason. *Complex TV: The Poetics of Contemporary Television Storytelling*. New York: New York University Press, 2015.

Modleski, Tania. *Loving Vengeance: Mass-Produced Fantasies for Women*. New York: Routledge, 1990.

Newman, Michael Z. "From Beats to Arcs: Toward a Poetics of Television Narrative." *Velvet Light Trap* 58, Fall 2006, pp. 16–28.

Newman, Michael Z., and Elana Levine. *Legitimating Television: Media Convergence and Cultural Status*. New York: Routledge, 2012.

Nussbaum, Emily. *I Like to Watch: Arguing My Way through the TV Revolution*. New York: Random House, 2020.

Och, Dana. "All Laura Palmer's Children: *Twin Peaks* and Gendering the Discourse of Influence." *Cinema Journal* 55 (3), 2016, pp. 131–136.

Odell, Jenny. *How to Do Nothing: Resisting the Attention Economy*. Brooklyn: Melville House, 2019.

Paella, Gabriella. "Hiding Out in the Wikipedia Page of a Horror Movie." *The Cut*. March 29, 2019. www.thecut.com.
Saxena, Jaya. "How to Watch Horror Movies if You're Too Scared of Horror Movies." *GQ.com*. December 7, 2017. www.gq.com.
Schramm, Wilbur, Jack Lyle, and Edwin B. Parker. *Television in the Lives of Our Children*. Stanford, CA: Stanford University Press, 1961.
Seitz, Matt Zoller. "The Sum and the Parts: In Defense of TV Recaps." *Vulture*. April 12, 2012. www.vulture.com.
Sepinwall, Alan. *The Cops, Crooks, Slingers, and Slayers Who Changed TV Drama Forever*. New York: Touchstone, 2012.
Shapiro, Jordan. *The New Childhood: Raising Kids to Thrive in a Connected World*. New York: Little, Brown Spark, 2018.
Smith-Rowsey, Daniel, and Kevin McDonald. *The Netflix Effect: Technology and Entertainment in the Twenty-First Century*. New York: Bloomsbury, 2016.
Spigel, Lynn. *Make Room for TV: Television and the Family Ideal in Postwar America*. Chicago: University of Chicago Press, 1992.
———. *Welcome to the Dream House: Popular Media and Postwar Suburbs*. Durham, NC: Duke University Press, 2001.
St. Felix, Doreen. "Black Vine: The Oral History of a Six-Second Movement." *MTV News*. November 4, 2016. www.mtv.com.
Starkman, Evan. "What Is Revenge Bedtime Procrastination?" *WebMD*. 2021. www.webmd.com.
Tichi, Cecelia. *Electronic Hearth: Creating an American Television Culture*. New York: Oxford University Press, 1991.
Tolentino, Jia. *Trick Mirror: Reflections on Self-Delusion*. New York: Random House, 2019.

Tryon, Chuck. *On-Demand Culture: Digital Delivery and the Future of Movies*. New Brunswick, NJ: Rutgers University Press, 2013.
Wallace, Carvell, host. *Finding Fred*. iHeart Radio, 2019.
Warner, Kristen. "Fans with Feels: *Game of Thrones* as Soap Opera." *Los Angeles Review of Books*. May 18, 2019. https://lareviewofbooks.org.
Wickman, Forrest. "Twin Peaks Didn't Just Change TV." *Slate*. May 3, 2017. https://slate.com.

About the Author

Phillip Maciak is the television editor at the *Los Angeles Review of Books*, and his essays have appeared in *Slate*, *The New Republic*, *n+1*, and other venues. He's the author of *The Disappearing Christ: Secularism in the Silent Era*. He teaches in the Department of English and the Program in American Culture Studies at Washington University in St. Louis.

Printed in the United States
by Baker & Taylor Publisher Services